Data Science Algorithms in a Week

Data analysis, machine learning, and more

Dávid Natingga

BIRMINGHAM - MUMBAI

Data Science Algorithms in a Week

First published: August 2017

Production reference: 1080817

Published by Packt Publishing Ltd.
Livery Place
35 Livery Street
Birmingham
B3 2PB, UK.

ISBN 978-1-78728-458-6

www.packtpub.com

Credits

Author
Dávid Natingga

Reviewer
Surendra Pepakayala

Commissioning Editor
Veena Pagare

Acquisition Editor
Chandan Kumar

Content Development Editor
Mamata Walkar

Technical Editor
Naveenkumar Jain

Copy Editor
Safis Editing

Project Coordinator
Kinjal Bari

Proofreader
Safis Editing

Indexer
Pratik Shirodkar

Production Coordinator
Shantanu Zagade

About the Author

Dávid Natingga graduated in 2014 from Imperial College London in MEng Computing with a specialization in Artificial Intelligence. In 2011, he worked at Infosys Labs in Bangalore, India, researching the optimization of machine learning algorithms. In 2012 and 2013, at Palantir Technologies in Palo Alto, USA, he developed algorithms for big data. In 2014, as a data scientist at Pact Coffee, London, UK, he created an algorithm suggesting products based on the taste preferences of customers and the structure of coffees. In 2017, he work at TomTom in Amsterdam, Netherlands, processing map data for navigation platforms.

As a part of his journey to use pure mathematics to advance the field of AI, he is a PhD candidate in Computability Theory at, University of Leeds, UK. In 2016, he spent 8 months at Japan, Advanced Institute of Science and Technology, Japan, as a research visitor.

Dávid Natingga married his wife Rheslyn and their first child will soon behold the outer world.

I would like to thank Packt Publishing for providing me with this opportunity to share my knowledge and experience in data science through this book. My gratitude belongs to my wife Rheslyn who has been patient, loving, and supportive through out the whole process of writing this book.

About the Reviewer

Surendra Pepakayala is a seasoned technology professional and entrepreneur with over 19 years of experience in the US and India. He has broad experience in building enterprise/web software products as a developer, architect, software engineering manager, and product manager at both start-ups and multinational companies in India and the US. He is a hands-on technologist/hacker with deep interest and expertise in Enterprise/Web Applications Development, Cloud Computing, Big Data, Data Science, Deep Learning, and Artificial Intelligence.

A technologist turned entrepreneur, after 11 years in corporate US, Surendra has founded an enterprise BI / DSS product for school districts in the US. He subsequently sold the company and started a Cloud Computing, Big Data, and Data Science consulting practice to help start-ups and IT organizations streamline their development efforts and reduce time to market of their products/solutions. Also, Surendra takes pride in using his considerable IT experience for reviving / turning-around distressed products / projects.

He serves as an advisor to eTeki, an on-demand interviewing platform, where he leads the effort to recruit and retain world-class IT professionals into eTeki's interviewer panel. He has reviewed drafts, recommended changes and formulated questions for various IT certifications such as CGEIT, CRISC, MSP, and TOGAF. His current focus is on applying Deep Learning to various stages of the recruiting process to help HR (staffing and corporate recruiters) find the best talent and reduce friction involved in the hiring process.

www.PacktPub.com

For support files and downloads related to your book, please visit www.PacktPub.com. Did you know that Packt offers eBook versions of every book published, with PDF and ePub files available? You can upgrade to the eBook version at www.PacktPub.com and as a print book customer, you are entitled to a discount on the eBook copy. Get in touch with us at service@packtpub.com for more details. At www.PacktPub.com, you can also read a collection of free technical articles, sign up for a range of free newsletters and receive exclusive discounts and offers on Packt books and eBooks.

https://www.packtpub.com/mapt

Get the most in-demand software skills with Mapt. Mapt gives you full access to all Packt books and video courses, as well as industry-leading tools to help you plan your personal development and advance your career.

Why subscribe?

- Fully searchable across every book published by Packt
- Copy and paste, print, and bookmark content
- On demand and accessible via a web browser

Customer Feedback

Thanks for purchasing this Packt book. At Packt, quality is at the heart of our editorial process. To help us improve, please leave us an honest review on this book's Amazon page at `link`.

If you'd like to join our team of regular reviewers, you can e-mail us at `customerreviews@packtpub.com`. We award our regular reviewers with free eBooks and videos in exchange for their valuable feedback. Help us be relentless in improving our products!

Table of Contents

Preface

Data science is a discipline at the intersection of machine learning, statistics and data mining with the objective to gain new knowledge from the existing data by the means of algorithmic and statistical analysis. In this book you will learn the 7 most important ways in Data Science to analyze the data. Each chapter first explains its algorithm or analysis as a simple concept supported by a trivial example. Further examples and exercises are used to build and expand the knowledge of a particular analysis.

What this book covers

`Chapter 1`, *Classification Using K Nearest Neighbors*, Classify a data item based on the k most similar items.

`Chapter 2`, *Naive Bayes*, Learn Bayes Theorem to compute the probability a data item belonging to a certain class.

`Chapter 3`, *Decision Trees*, Organize your decision criteria into the branches of a tree and use a decision tree to classify a data item into one of the classes at the leaf node.

`Chapter 4`, *Random Forest*, Classify a data item with an ensemble of decision trees to improve the accuracy of the algorithm by reducing the negative impact of the bias.

`Chapter 5`, *Clustering into K Clusters*, Divide your data into k clusters to discover the patterns and similarities between the data items. Exploit these patterns to classify new data.

`Chapter 6`, *Regression*, Model a phenomena in your data by a function that can predict the values for the unknown data in a simple way.

`Chapter 7`, *Time Series Analysis*, Unveil the trend and repeating patters in time dependent data to predict the future of the stock market, Bitcoin prices and other time events.

`Appendix A`, *Statistics*, Provides a summary of the statistical methods and tools useful to a data scientist.

`Appendix B`, *R Reference*, Reference to the basic Basic Python language constructs; libraries Numpy and Pandas used throughout the book.

`Appendix C`, *Python Reference*, Reference to the basic Basic Basic R language constructs, commands and functions used throughout the book.

`Appendix D`, *Glossary of Algorithms and Methods in Data Science,* Provides a glossary for some of the most important and powerful algorithms and methods from the fields of the data science and machine learning.

What you need for this book

Most importantly, an active attitude to think of the problems--a lot of new content is presented in the exercises. Then you also need to be able to run Python and R programs under the operating system of your choice. The author ran the programs under Linux operating system using command line.

Who this book is for

This book is for aspiring data science professionals who are familiar with Python & R and have some statistics background. Those developers who are currently implementing 1 or 2 data science algorithms and now want to learn more to expand their skill will find this book quite useful.

Conventions

In this book, you will find a number of text styles that distinguish between different kinds of information. Here are some examples of these styles and an explanation of their meaning. Code words in text, database table names, folder names, filenames, file extensions, pathnames, dummy URLs, user input, and Twitter handles are shown as follows: "For the visualization depicted earlier in this chapter, the `matplotlib` library was used."

A block of code is set as follows:

```
import sys
sys.path.append('..')
sys.path.append('../../common')
import knn # noqa
import common # noqa
```

Any command-line input or output is written as follows:

```
$ python knn_to_data.py mary_and_temperature_preferences.data
mary_and_temperature_preferences_completed.data 1 5 30 0 10
```

New terms and **important words** are shown in bold. Words that you see on the screen, for example, in menus or dialog boxes, appear in the text like this: "In order to download new modules, we will go to **Files** | **Settings** | **Project Name** | **Project Interpreter**."

Warnings or important notes appear like this.

Tips and tricks appear like this.

Reader feedback

Feedback from our readers is always welcome. Let us know what you think about this book-what you liked or disliked. Reader feedback is important for us as it helps us develop titles that you will really get the most out of. To send us general feedback, simply e-mail `feedback@packtpub.com`, and mention the book's title in the subject of your message. If there is a topic that you have expertise in and you are interested in either writing or contributing to a book, see our author guide at `www.packtpub.com/authors`.

Customer support

Now that you are the proud owner of a Packt book, we have a number of things to help you to get the most from your purchase.

Downloading the example code

You can download the example code files for this book from your account at `http://www.packtpub.com`. If you purchased this book elsewhere, you can visit `http://www.packtpub.com/support`and register to have the files e-mailed directly to you. You can download the code files by following these steps:

1. Log in or register to our website using your e-mail address and password.
2. Hover the mouse pointer on the **SUPPORT** tab at the top.
3. Click on **Code Downloads & Errata**.
4. Enter the name of the book in the **Search** box.
5. Select the book for which you're looking to download the code files.
6. Choose from the drop-down menu where you purchased this book from.
7. Click on **Code Download**.

Once the file is downloaded, please make sure that you unzip or extract the folder using the latest version of:

- WinRAR / 7-Zip for Windows
- Zipeg / iZip / UnRarX for Mac
- 7-Zip / PeaZip for Linux

The code bundle for the book is also hosted on GitHub at `https://github.com/PacktPublishing/Data-Science-Algorithms-in-a-Week`. We also have other code bundles from our rich catalog of books and videos available at `https://github.com/PacktPublishing/`. Check them out!

Downloading the color images of this book

We also provide you with a PDF file that has color images of the screenshots/diagrams used in this book. The color images will help you better understand the changes in the output. You can download this file from `https://www.packtpub.com/sites/default/files/downloads/DataScienceAlgorithmsinaWeek_ColorImages.pdf`.

Errata

Although we have taken every care to ensure the accuracy of our content, mistakes do happen. If you find a mistake in one of our books-maybe a mistake in the text or the code-we would be grateful if you could report this to us. By doing so, you can save other readers from frustration and help us improve subsequent versions of this book. If you find any errata, please report them by visiting `http://www.packtpub.com/submit-errata`, selecting your book, clicking on the **Errata Submission Form** link, and entering the details of your errata. Once your errata are verified, your submission will be accepted and the errata will be uploaded to our website or added to any list of existing errata under the Errata section of that title. To view the previously submitted errata, go to `https://www.packtpub.com/books/content/support`and enter the name of the book in the search field. The required information will appear under the **Errata** section.

Piracy

Piracy of copyrighted material on the Internet is an ongoing problem across all media. At Packt, we take the protection of our copyright and licenses very seriously. If you come across any illegal copies of our works in any form on the Internet, please provide us with the location address or website name immediately so that we can pursue a remedy. Please contact us at `copyright@packtpub.com` with a link to the suspected pirated material. We appreciate your help in protecting our authors and our ability to bring you valuable content.

Questions

If you have a problem with any aspect of this book, you can contact us at `questions@packtpub.com`, and we will do our best to address the problem.

1
Classification Using K Nearest Neighbors

The nearest neighbor algorithm classifies a data instance based on its neighbors. The class of a data instance determined by the k-nearest neighbor algorithm is the class with the highest representation among the k-closest neighbors.

In this chapter, we will cover the basics of the k-NN algorithm - understanding it and its implementation with a simple example: Mary and her temperature preferences. On the example map of Italy, you will learn how to choose a correct value *k* so that the algorithm can perform correctly and with the highest accuracy. You will learn how to rescale the values and prepare them for the k-NN algorithm with the example of house preferences. In the example of text classification, you will learn how to choose a good metric to measure the distances between the data points, and also how to eliminate the irrelevant dimensions in higher-dimensional space to ensure that the algorithm performs accurately.

Mary and her temperature preferences

As an example, if we know that our friend Mary feels cold when it is 10 degrees Celsius, but warm when it is 25 degrees Celsius, then in a room where it is 22 degrees Celsius, the nearest neighbor algorithm would guess that our friend would feel warm, because 22 is closer to 25 than to 10.

Suppose we would like to know when Mary feels warm and when she feels cold, as in the previous example, but in addition, wind speed data is also available when Mary was asked if she felt warm or cold:

Temperature in degrees Celsius	Wind speed in km/h	Mary's perception
10	0	Cold
25	0	Warm
15	5	Cold
20	3	Warm
18	7	Cold
20	10	Cold
22	5	Warm
24	6	Warm

We could represent the data in a graph, as follows:

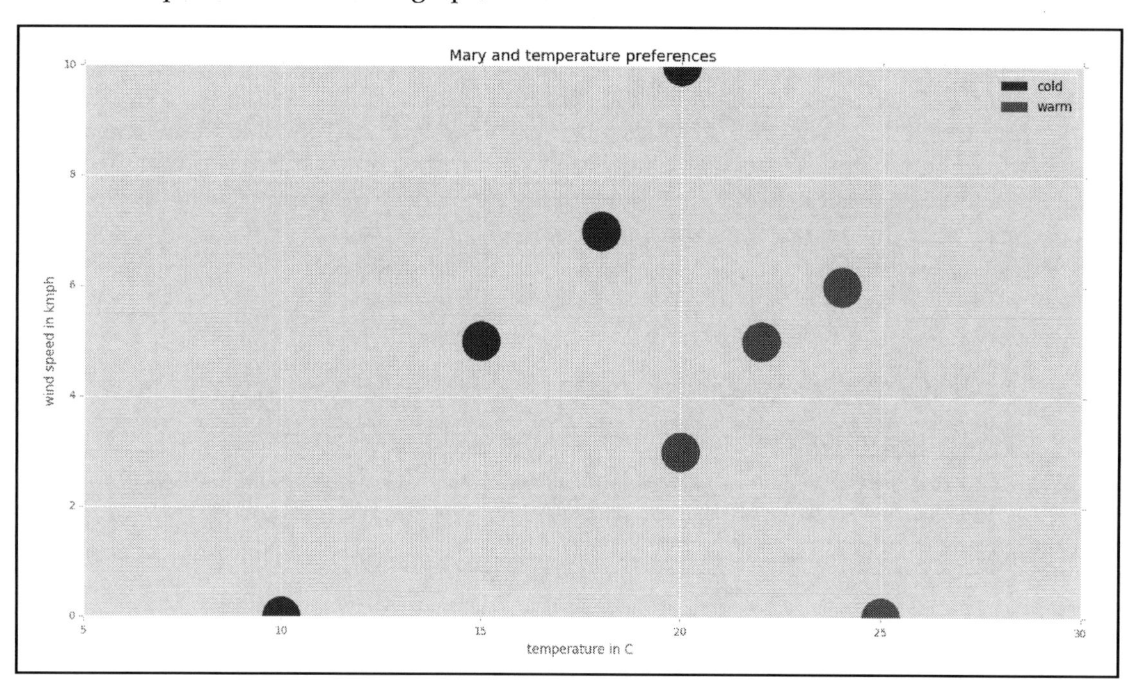

Now, suppose we would like to find out how Mary feels at the temperature 16 degrees Celsius with a wind speed of 3km/h using the 1-NN algorithm:

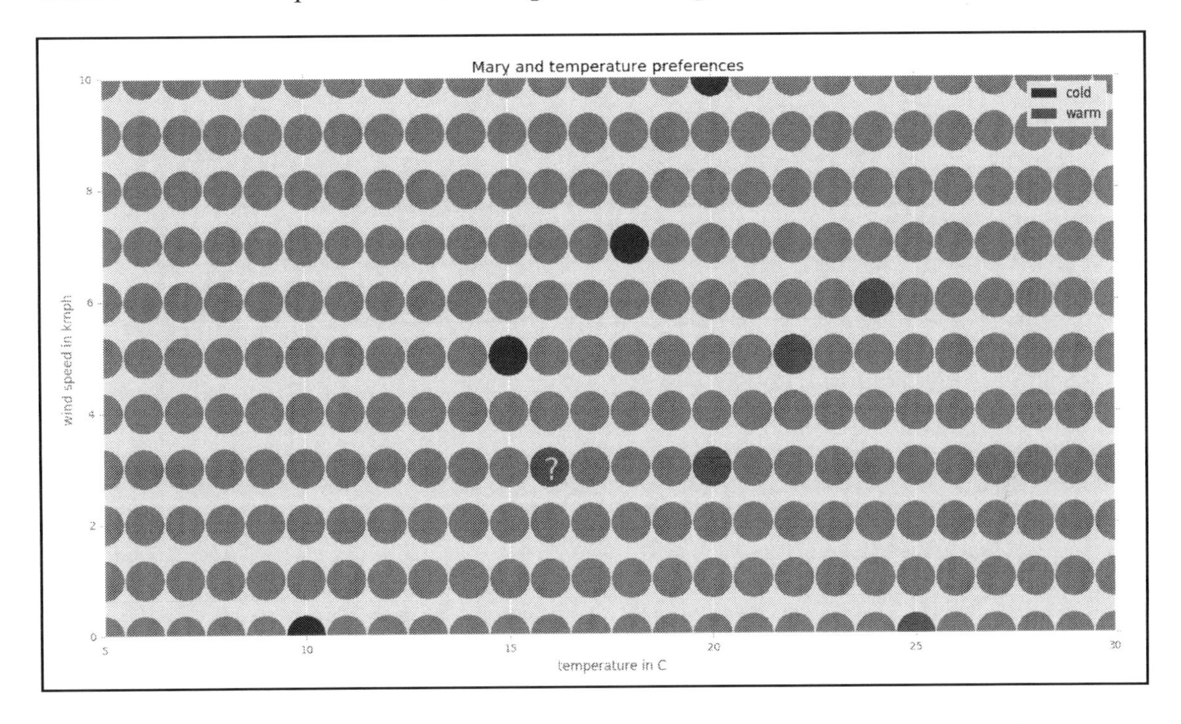

For simplicity, we will use a Manhattan metric to measure the distance between the neighbors on the grid. The Manhattan distance d_{Man} of a neighbor $N_1=(x_1,y_1)$ from the neighbor $N_2=(x_2,y_2)$ is defined to be $d_{Man}=|x_1- x_2|+|y_1- y_2|$.

Let us label the grid with distances around the neighbors to see which neighbor with a known class is closest to the point we would like to classify:

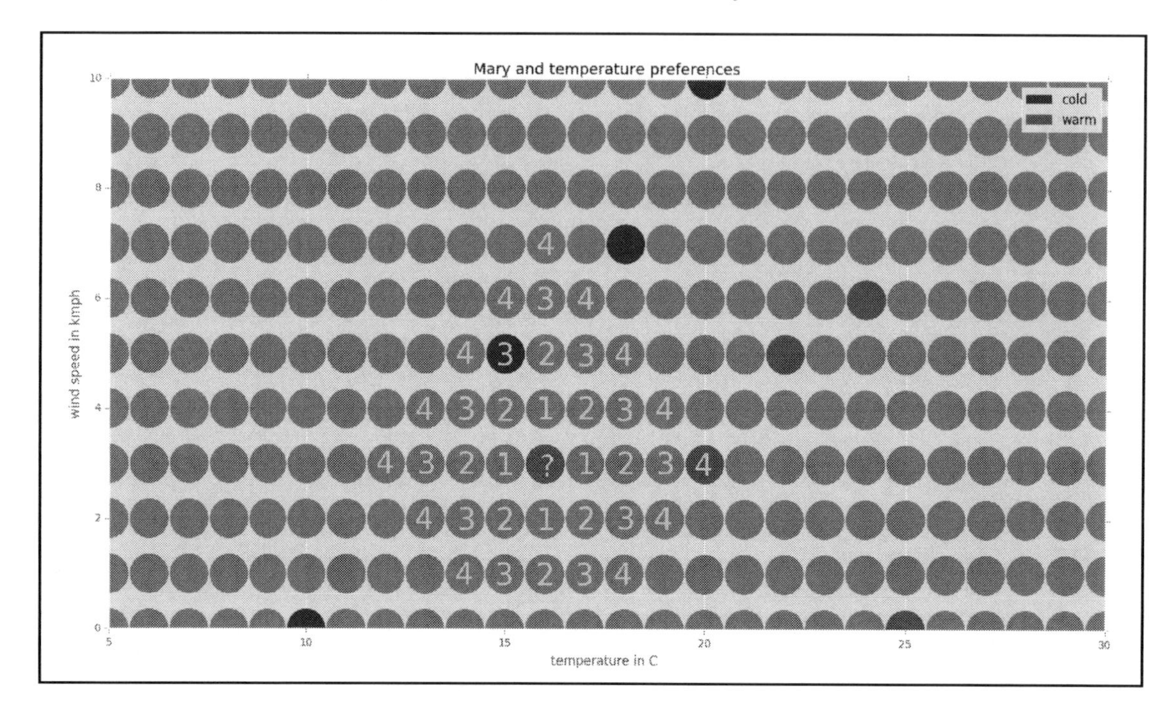

We can see that the closest neighbor with a known class is the one with the temperature 15 (blue) degrees Celsius and the wind speed 5km/h. Its distance from the questioned point is three units. Its class is blue (cold). The closest red (warm) neighbor is distanced four units from the questioned point. Since we are using the 1-nearest neighbor algorithm, we just look at the closest neighbor and, therefore, the class of the questioned point should be blue (cold).

By applying this procedure to every data point, we can complete the graph as follows:

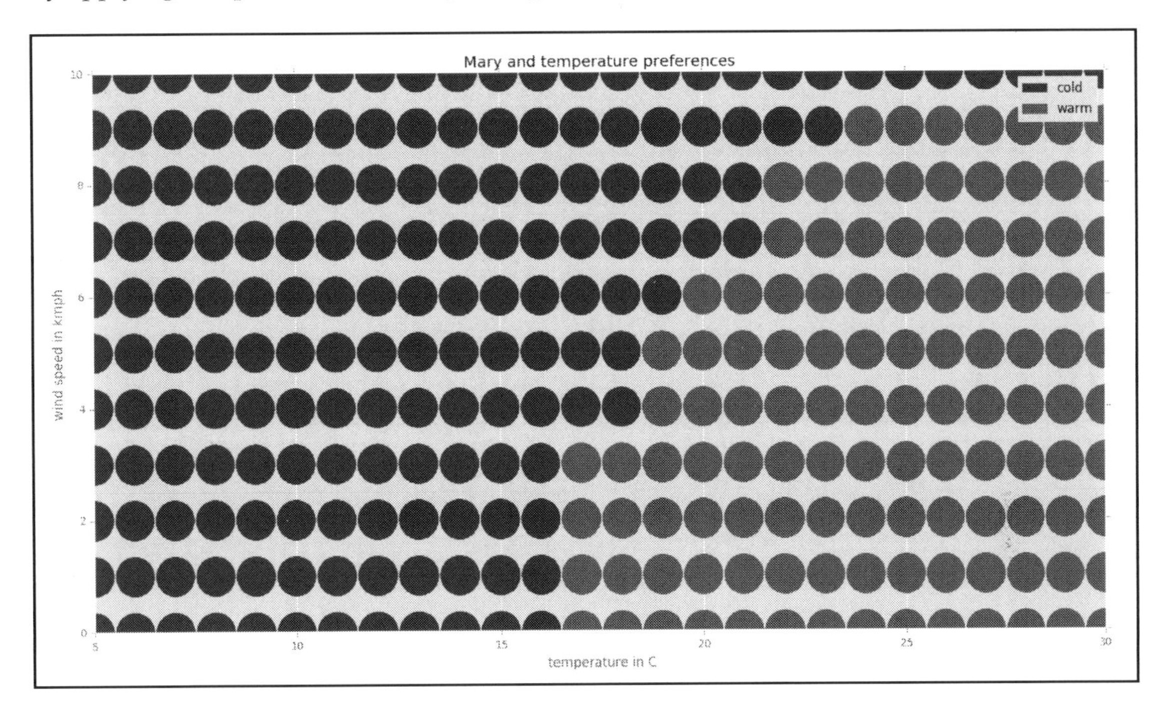

Note that sometimes a data point can be distanced from two known classes with the same distance: for example, 20 degrees Celsius and 6km/h. In such situations, we could prefer one class over the other or ignore these boundary cases. The actual result depends on the specific implementation of an algorithm.

Implementation of k-nearest neighbors algorithm

We implement the k-NN algorithm in Python to find Mary's temperature preference. In the end of this section we also implement the visualization of the data produced in example Mary and her temperature preferences by the k-NN algorithm. The full compilable code with the input files can be found in the source code provided with this book. The most important parts are extracted here:

```
# source_code/1/mary_and_temperature_preferences/knn_to_data.py
# Applies the knn algorithm to the input data.
```

```
# The input text file is assumed to be of the format with one line per
# every data entry consisting of the temperature in degrees Celsius,
# wind speed and then the classification cold/warm.

import sys
sys.path.append('..')
sys.path.append('../../common')
import knn # noqa
import common # noqa

# Program start
# E.g. "mary_and_temperature_preferences.data"
input_file = sys.argv[1]
# E.g. "mary_and_temperature_preferences_completed.data"
output_file = sys.argv[2]
k = int(sys.argv[3])
x_from = int(sys.argv[4])
x_to = int(sys.argv[5])
y_from = int(sys.argv[6])
y_to = int(sys.argv[7])

data = common.load_3row_data_to_dic(input_file)
new_data = knn.knn_to_2d_data(data, x_from, x_to, y_from, y_to, k)
common.save_3row_data_from_dic(output_file, new_data)

# source_code/common/common.py
# ***Library with common routines and functions***
def dic_inc(dic, key):
    if key is None:
        pass
    if dic.get(key, None) is None:
        dic[key] = 1
    else:
        dic[key] = dic[key] + 1

# source_code/1/knn.py
# ***Library implementing knn algorihtm***

def info_reset(info):
    info['nbhd_count'] = 0
    info['class_count'] = {}

# Find the class of a neighbor with the coordinates x,y.
# If the class is known count that neighbor.
def info_add(info, data, x, y):
    group = data.get((x, y), None)
    common.dic_inc(info['class_count'], group)
    info['nbhd_count'] += int(group is not None)
```

```
# Apply knn algorithm to the 2d data using the k-nearest neighbors with
# the Manhattan distance.
# The dictionary data comes in the form with keys being 2d coordinates
# and the values being the class.
# x,y are integer coordinates for the 2d data with the range
# [x_from,x_to] x [y_from,y_to].
def knn_to_2d_data(data, x_from, x_to, y_from, y_to, k):
    new_data = {}
    info = {}
    # Go through every point in an integer coordinate system.
    for y in range(y_from, y_to + 1):
        for x in range(x_from, x_to + 1):
            info_reset(info)
            # Count the number of neighbors for each class group for
            # every distance dist starting at 0 until at least k
            # neighbors with known classes are found.
            for dist in range(0, x_to - x_from + y_to - y_from):
                # Count all neighbors that are distanced dist from
                # the point [x,y].
                if dist == 0:
                    info_add(info, data, x, y)
                else:
                    for i in range(0, dist + 1):
                        info_add(info, data, x - i, y + dist - i)
                        info_add(info, data, x + dist - i, y - i)
                    for i in range(1, dist):
                        info_add(info, data, x + i, y + dist - i)
                        info_add(info, data, x - dist + i, y - i)
                # There could be more than k-closest neighbors if the
                # distance of more of them is the same from the point
                # [x,y]. But immediately when we have at least k of
                # them, we break from the loop.
                if info['nbhd_count'] >= k:
                    break
            class_max_count = None
            # Choose the class with the highest count of the neighbors
            # from among the k-closest neighbors.
            for group, count in info['class_count'].items():
                if group is not None and (class_max_count is None or
                   count > info['class_count'][class_max_count]):
                    class_max_count = group
            new_data[x, y] = class_max_count
    return new_data
```

Input:

The program above will use the file below as the source of the input data. The file contains the table with the known data about Mary's temperature preferences:

```
# source_code/1/mary_and_temperature_preferences/
marry_and_temperature_preferences.data
10 0 cold
25 0 warm
15 5 cold
20 3 warm
18 7 cold
20 10 cold
22 5 warm
24 6 warm
```

Output:

We run the implementation above on the input file `mary_and_temperature_preferences.data` using the `k-NN` algorithm for `k=1` neighbors. The algorithm classifies all the points with the integer coordinates in the rectangle with a size of `(30-5=25) by (10-0=10)`, so with the a of `(25+1) * (10+1) = 286` integer points (adding one to count points on boundaries). Using the `wc` command, we find out that the output file contains exactly 286 lines - one data item per point. Using the head command, we display the first 10 lines from the output file. We visualize all the data from the output file in the next section:

```
$ python knn_to_data.py mary_and_temperature_preferences.data
mary_and_temperature_preferences_completed.data 1 5 30 0 10

$ wc -l mary_and_temperature_preferences_completed.data
286 mary_and_temperature_preferences_completed.data

$ head -10 mary_and_temperature_preferences_completed.data
7 3 cold
6 9 cold
12 1 cold
16 6 cold
16 9 cold
14 4 cold
13 4 cold
19 4 warm
18 4 cold
15 1 cold
```

Visualization:

For the visualization depicted earlier in this chapter, the `matplotlib` library was used. A data file is loaded, and then displayed in a scattered diagram:

```
# source_code/common/common.py
# returns a dictionary of 3 lists: 1st with x coordinates,
# 2nd with y coordinates, 3rd with colors with numeric values
def get_x_y_colors(data):
    dic = {}
    dic['x'] = [0] * len(data)
    dic['y'] = [0] * len(data)
    dic['colors'] = [0] * len(data)
    for i in range(0, len(data)):
        dic['x'][i] = data[i][0]
        dic['y'][i] = data[i][1]
        dic['colors'][i] = data[i][2]
    return dic
```

```
# source_code/1/mary_and_temperature_preferences/
mary_and_temperature_preferences_draw_graph.py
import sys
sys.path.append('../../common')  # noqa
import common
import numpy as np
import matplotlib.pyplot as plt
import matplotlib.patches as mpatches
import matplotlib
matplotlib.style.use('ggplot')

data_file_name = 'mary_and_temperature_preferences_completed.data'
temp_from = 5
temp_to = 30
wind_from = 0
wind_to = 10

data = np.loadtxt(open(data_file_name, 'r'),
                  dtype={'names': ('temperature', 'wind', 'perception'),
                         'formats': ('i4', 'i4', 'S4')})

# Convert the classes to the colors to be displayed in a diagram.
for i in range(0, len(data)):
    if data[i][2] == 'cold':
        data[i][2] = 'blue'
    elif data[i][2] == 'warm':
        data[i][2] = 'red'
    else:
        data[i][2] = 'gray'
```

```
# Convert the array into the format ready for drawing functions.
data_processed = common.get_x_y_colors(data)

# Draw the graph.
plt.title('Mary and temperature preferences')
plt.xlabel('temperature in C')
plt.ylabel('wind speed in kmph')
plt.axis([temp_from, temp_to, wind_from, wind_to])
# Add legends to the graph.
blue_patch = mpatches.Patch(color='blue', label='cold')
red_patch = mpatches.Patch(color='red', label='warm')
plt.legend(handles=[blue_patch, red_patch])
plt.scatter(data_processed['x'], data_processed['y'],
            c=data_processed['colors'], s=[1400] * len(data))
plt.show()
```

Map of Italy example - choosing the value of k

In our data, we are given some points (about 1%) from the map of Italy and its surroundings. The blue points represent water and the green points represent land; white points are not known. From the partial information given, we would like to predict whether there is water or land in the white areas.

Drawing only 1% of the map data in the picture would be almost invisible. If, instead, we were given about 33 times more data from the map of Italy and its surroundings and drew it in the picture, it would look like below:

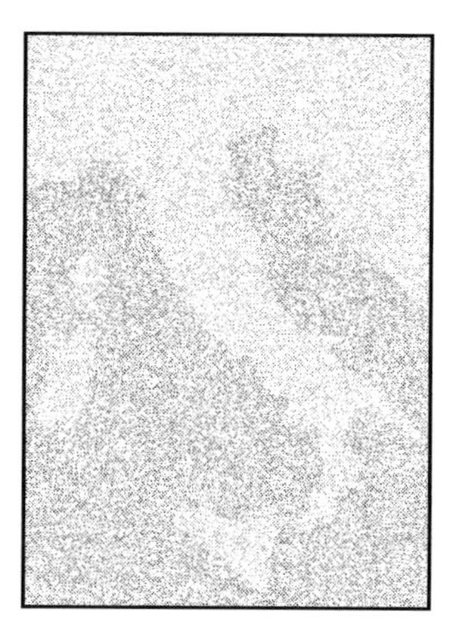

Analysis:

For this problem, we will use the k-NN algorithm - k here means that we will look at *k* closest neighbors. Given a white point, it will be classified as a water area if the majority of its *k* closest neighbors are in the water area, and classified as land if the majority of its *k* closest neighbors are in the land area. We will use the Euclidean metric for the distance: given two points $X=[x_0,x_1]$ and $Y=[y_0,y_1]$, their Euclidean distance is defined as $d_{Euclidean} = sqrt((x_0-y_0)^2+(x_1-y_1)^2)$.

The Euclidean distance is the most common metric. Given two points on a piece of paper, their Euclidean distance is just the length between the two points, as measured by a ruler, as shown in the diagram:

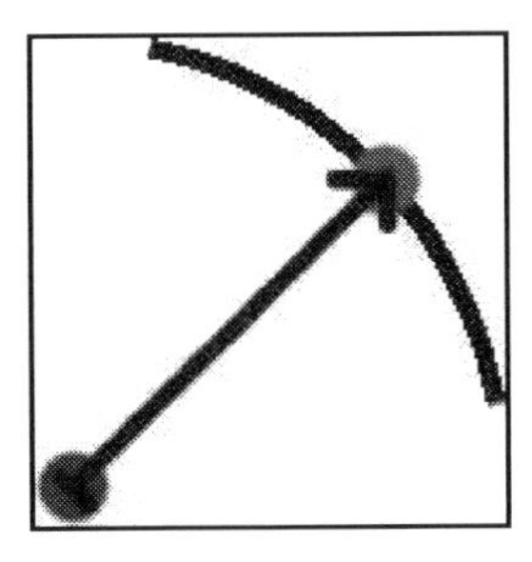

To apply the k-NN algorithm to an incomplete map, we have to choose the value of k. Since the resulting class of a point is the class of the majority of the *k* closest neighbors of that point, *k* should be odd. Let us apply the algorithm for the values of *k=1,3,5,7,9*.

Applying this algorithm to every white point of the incomplete map will result in the following completed maps:

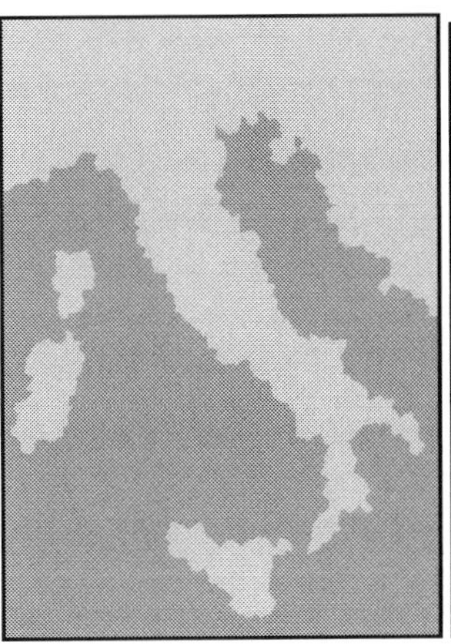

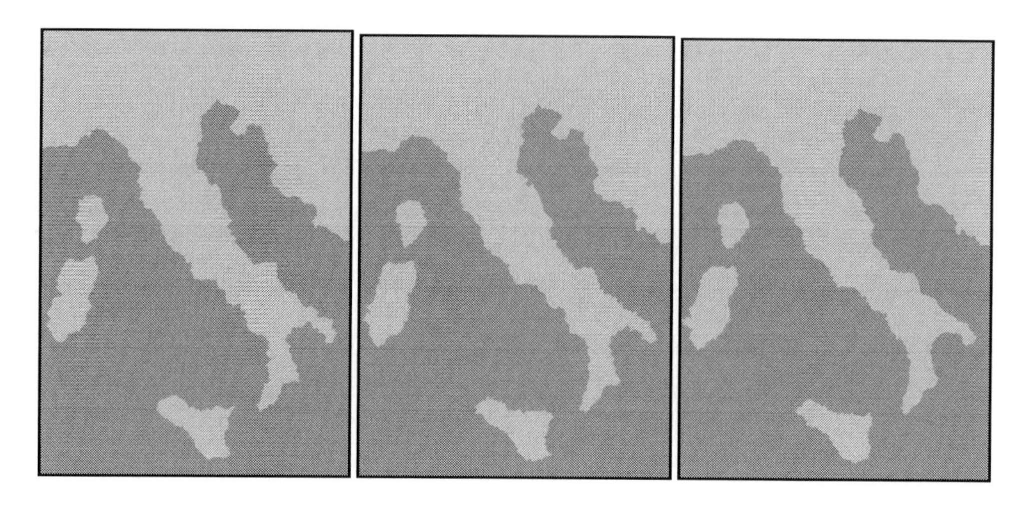

As you will notice, the higher value of *k* results in a completed map with smoother boundaries. The actual complete map of Italy is here:

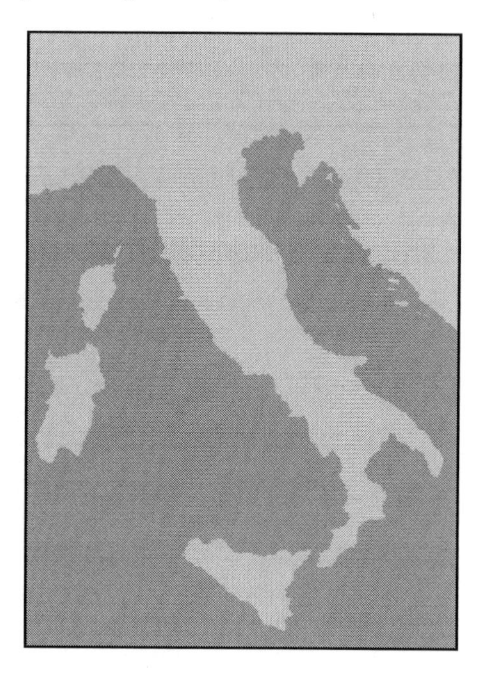

We can use this real completed map to calculate the percentage of the incorrectly classified points for the various values of *k* to determine the accuracy of the k-NN algorithm for different values of k:

k	% of incorrectly classified points
1	2.97
3	3.24
5	3.29
7	3.40
9	3.57

Thus, for this particular type of classification problem, the k-NN algorithm achieves the highest accuracy (least error rate) for *k=1*.

However, in real-life, problems we wouldn't usually not have complete data or a solution. In such scenarios, we need to choose *k* appropriate to the partially available data. For this, consult problem 1.4.

House ownership - data rescaling

For each person, we are given their age, yearly income, and whether their is a house or not:

Age	Annual income in USD	House ownership status
23	50,000	Non-owner
37	34,000	Non-owner
48	40,000	Owner
52	30,000	Non-owner
28	95,000	Owner
25	78,000	Non-owner
35	130,000	Owner
32	105,000	Owner
20	100,000	Non-owner
40	60,000	Owner
50	80,000	Peter

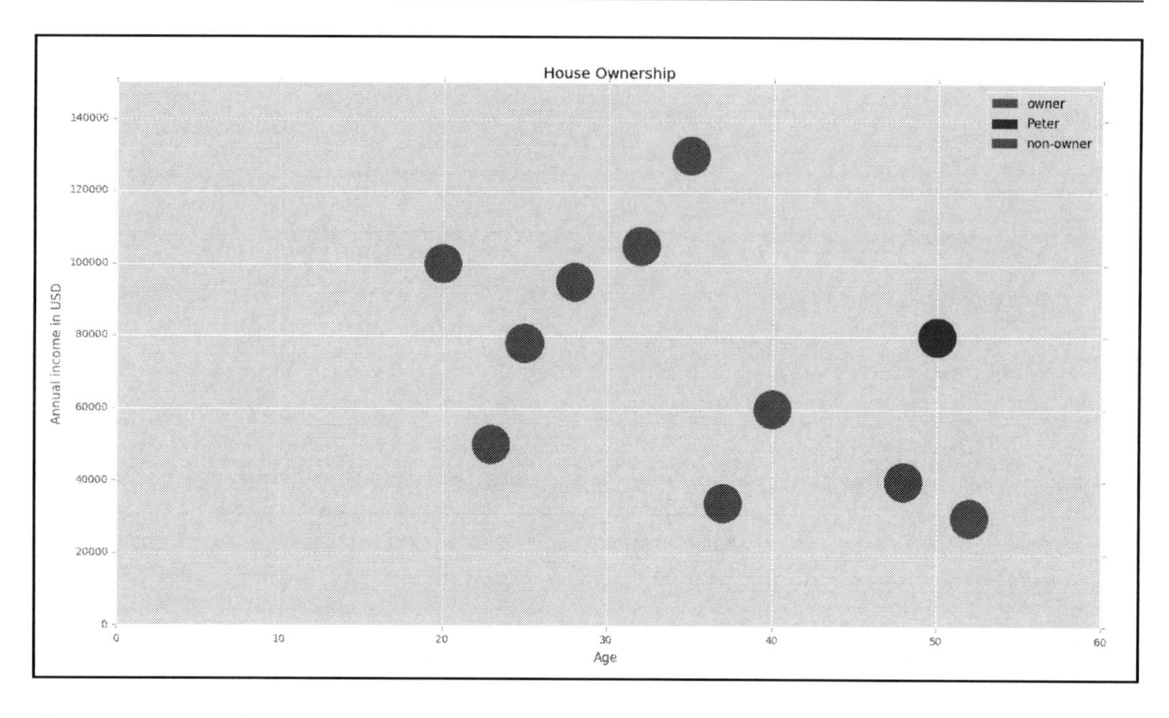

The aim is to predict whether Peter, aged 50, with an income of $80k/year, owns a house and could be a potential customer for our insurance company.

Analysis:

In this case, we could try to apply the 1-NN algorithm. However, we should be careful about how we are going to measure the distances between the data points, since the income range is much wider than the age range. Income levels of $115k and $116k are $1,000 apart. These two data points with these incomes would result in a very long distance. However, relative to each other, the difference is not too large. Because we consider both measures (age and yearly income) to be about as important, we would scale both from 0 to 1 according to the formula:

ScaledQuantity = (ActualQuantity-MinQuantity)/(MaxQuantity-MinQuantity)

In our particular case, this reduces to:

ScaledAge = (ActualAge-MinAge)/(MaxAge-MinAge)

ScaledIncome = (ActualIncome- inIncome)/(MaxIncome-inIncome)

After scaling, we get the following data:

Age	Scaled age	Annual income in USD	Scaled annual income	House ownership status
23	0.09375	50,000	0.2	Non-owner
37	0.53125	34,000	0.04	Non-owner
48	0.875	40,000	0.1	Owner
52	1	30,000	0	Non-owner
28	0.25	95,000	0.65	Owner
25	0.15625	78,000	0.48	Non-owner
35	0.46875	130,000	1	Owner
32	0.375	105,000	0.75	Owner
20	0	100,000	0.7	Non-owner
40	0.625	60,000	0.3	Owner
50	0.9375	80,000	0.5	?

Now, if we apply the 1-NN algorithm with the Euclidean metric, we will find out that Peter more than likely owns a house. Note that, without rescaling, the algorithm would yield a different result. Refer to exercise 1.5.

Text classification - using non-Euclidean distances

We are given the word counts of the keywords **algorithm** and **computer** for documents of the classes, informatics and mathematics:

Algorithm words per 1,000	Computer words per 1,000	Subject classification
153	150	Informatics
105	97	Informatics
75	125	Informatics
81	84	Informatics

73	77	Informatics
90	63	Informatics
20	0	Mathematics
33	0	Mathematics
105	10	Mathematics
2	0	Mathematics
84	2	Mathematics
12	0	Mathematics
41	42	?

The documents with a high rate of the words **algorithm** and **computer** are in the class of `informatics`. The class of mathematics happens to contain documents with a high count of the word **algorithm** in some cases; for example, a document concerned with the Euclidean algorithm from the field of number theory. But, since mathematics tends to be less applied than `informatics` in the area of algorithms, the word **computer** is contained in such documents with a lower frequency.

We would like to classify a document that has 41 instances of the word **algorithm** per 1,000 words and 42 instances of the word **computer** per 1,000 words:

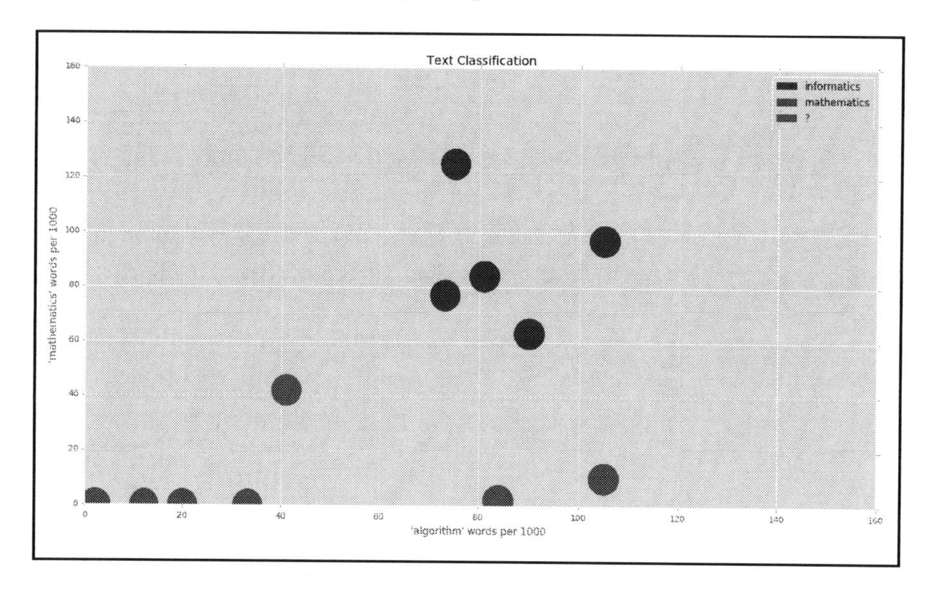

Analysis:

Using, for example, the 1-NN algorithm and the Manhattan or Euclidean distance would result in the classification of the document in question to the class of mathematics. However, intuitively, we should instead use a different metric to measure the distance, as the document in question has a much higher count of the word **computer** than other known documents in the class of mathematics.

Another candidate metric for this problem is a metric that would measure the proportion of the counts for the words, or the angle between the instances of documents. Instead of the angle, one could take the cosine of the angle $cos(\theta)$, and then use the well-known dot product formula to calculate the $cos(\theta)$.

Let $a=(a_x,a_y)$, $b=(b_x,b_y)$, then instead this formula:

$$|a||b|\cos(\theta) = a \cdot b = a_x \cdot b_x + a_y \cdot b_y$$

One derives:

$$cos(\theta) = \frac{a_x \cdot b_x + a_y \cdot b_y}{|a||b|}$$

Using the cosine distance metric, one could classify the document in question to the class of informatics:

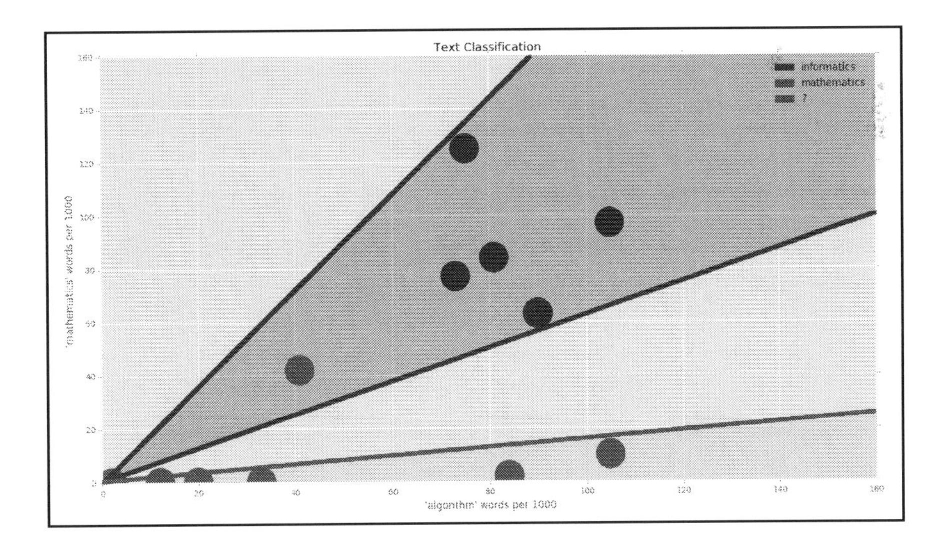

Text classification - k-NN in higher-dimensions

Suppose we are given documents and we would like to classify other documents based on their word frequency counts. For example, the 120 most frequent words for the Project Gutenberg e-book of the King James Bible are as follows:

1. the 8.07%	41. when 0.36%	81. go 0.19%
2. and 6.51%	42. this 0.36%	82. hand 0.18%
3. of 4.37%	43. out 0.35%	83. us 0.18%
4. to 1.72%	44. were 0.35%	84. saying 0.18%
5. that 1.63%	45. upon 0.35%	85. made 0.18%
6. in 1.60%	46. man 0.34%	87. went 0.18%
7. he 1.31%	47. you 0.34%	88. even 0.18%
8. shall 1.24%	48. by 0.33%	89. do 0.17%
9. for 1.13%	49. Israel 0.32%	90. now 0.17%
10. unto 1.13%	50. king 0.32%	91. behold 0.17%
11. i 1.11%	51. son 0.30%	92. saith 0.16%
12. his 1.07%	52. up 0.30%	93. therefore 0.16%
13. a 1.04%	53. there 0.29%	94. every 0.16%
14. lord 1.00%	54. hath 0.28%	95. these 0.15%
15. they 0.93%	55. then 0.27%	96. because 0.15%
16. be 0.88%	56. people 0.27%	97. or 0.15%
17. is 0.88%	57. came 0.26%	98. after 0.15%
18. him 0.84%	58. had 0.25%	99. our 0.15%
19. not 0.83%	59. house 0.25%	100. things 0.15%
20. them 0.81%	60. on 0.25%	101. father 0.14%
21. it 0.77%	61. into 0.25%	102. down 0.14%
22. with 0.76%	62. her 0.25%	103. sons 0.14%
23. all 0.71%	63. come 0.25%	104. hast 0.13%
24. thou 0.69%	64. one 0.25%	105. David 0.13%
25. thy 0.58%	65. we 0.23%	106. o 0.13%
26. was 0.57%	66. children 0.23%	107. make 0.13%
27. god 0.56%	67. s 0.23%	108. say 0.13%
28. which 0.56%	68. before 0.23%	109. may 0.13%
29. my 0.55%	69. your 0.23%	110. over 0.13%
30. me 0.52%	70. also 0.22%	111. did 0.13%
31. said 0.50%	71. day 0.22%	112. earth 0.12%
32. but 0.50%	72. land 0.22%	113. what 0.12%
33. ye 0.50%	74. so 0.21%	114. Jesus 0.12%
34. their 0.50%	75. men 0.21%	115. she 0.12%
35. have 0.49%	76. against 0.21%	116. who 0.12%
36. will 0.48%	77. shalt 0.20%	117. great 0.12%
37. thee 0.48%	78. if 0.20%	118. name 0.12%
38. from 0.46%	79. at 0.20%	119. any 0.12%
39. as 0.44%	80. let 0.19%	120. thine 0.12%
40. are 0.37%		

The task is to design a metric which, given the word frequencies for each document, would accurately determine how semantically close those documents are. Consequently, such a metric could be used by the k-NN algorithm to classify the unknown instances of the new documents based on the existing documents.

Analysis:

Suppose that we consider, for example, N most frequent words in our corpus of the documents. Then, we count the word frequencies for each of the N words in a given document and put them in an N dimensional vector that will represent that document. Then, we define a distance between two documents to be the distance (for example, Euclidean) between the two word frequency vectors of those documents.

The problem with this solution is that only certain words represent the actual content of the book, and others need to be present in the text because of grammar rules or their general basic meaning. For example, out of the 120 most frequent words in the Bible, each word is of a different importance, and the author highlighted the words in bold that have an especially high frequency in the Bible and bear an important meaning:

• lord - used 1.00% • god - 0.56%	• Israel - 0.32% • king - 0.32%	• David - 0.13% • Jesus - 0.12%

These words are less likely to be present in the mathematical texts for example, but more likely to be present in the texts concerned with religion or Christianity.

However, if we just look at the six most frequent words in the Bible, they happen to be less in detecting the meaning of the text:

• the 8.07% • and 6.51%	• of 4.37% • to 1.72%	• that 1.63% • in 1.60%

Texts concerned with mathematics, literature, or other subjects will have similar frequencies for these words. The differences may result mostly from the writing style.

Therefore, to determine a similarity distance between two documents, we need to look only at the frequency counts of the important words. Some words are less important - these dimensions are better reduced, as their inclusion can lead to a misinterpretation of the results in the end. Thus, what we are left to do is to choose the words (dimensions) that are important to classify the documents in our corpus. For this, consult exercise 1.6.

Summary

The k-nearest neighbor algorithm is a classification algorithm that assigns to a given data point the majority class among the k-nearest neighbors. The distance between two points is measured by a metric. Examples of distances include: Euclidean distance, Manhattan distance, Minkowski distance, Hamming distance, Mahalanobis distance, Tanimoto distance, Jaccard distance, tangential distance, and cosine distance. Experiments with various parameters and cross-validation can help to establish which parameter *k* and which metric should be used.

The dimensionality and position of a data point in the space are determined by its qualities. A large number of dimensions can result in low accuracy of the k-NN algorithm. Reducing the dimensions of qualities of smaller importance can increase accuracy. Similarly, to increase accuracy further, distances for each dimension should be scaled according to the importance of the quality of that dimension.

Problems

1. **Mary and her temperature preferences**: Imagine that you know that your friend Mary feels cold when it is -50 degrees Celsius, but she feels warm when it is 20 degrees Celsius. What would the 1-NN algorithm say about Mary; would she feel warm or cold at the temperatures 22, 15, -10? Do you think that the algorithm predicted Mary's body perception of the temperature correctly? If not, please, give the reasons and suggest why the algorithm did not give appropriate results and what would need to improve in order for the algorithm to make a better classification.
2. **Mary and temperature preferences**: Do you think that the use of the 1-NN algorithm would yield better results than the use of the k-NN algorithm for k>1?
3. **Mary and temperature preferences**: We collected more data and found out that Mary feels warm at 17C, but cold at 18C. By our common sense, Mary should feel warmer with a higher temperature. Can you explain a possible cause of discrepancy in the data? How could we improve the analysis of our data? Should we collect also some non-temperature data? Suppose that we have only temperature data available, do you think that the 1-NN algorithm would still yield better results with the data like this? How should we choose *k* for k-NN algorithm to perform well?

4. **Map of Italy - choosing the value of k**: We are given a partial map of Italy as for the problem Map of Italy. But suppose that the complete data is not available. Thus we cannot calculate the error rate on all the predicted points for different values of k. How should one choose the value of *k* for the k-NN algorithm to complete the map of Italy in order to maximize the accuracy?
5. **House ownership**: Using the data from the section concerned with the problem of house ownership, find the closest neighbor to Peter using the Euclidean metric:
    1. without rescaling the data,
    2. using the scaled data.

    Is the closest neighbor in a) the same as the neighbor in b)? Which of the neighbors owns the house?

6. **Text classification**: Suppose you would like to find books or documents in Gutenberg's corpus (www.gutenberg.org) that are similar to a selected book from the corpus (for example, the Bible) using a certain metric and the 1-NN algorithm. How would you design a metric measuring the similarity distance between the two documents?

Analysis:

1. 8 degrees Celsius is closer to 20 degrees Celsius than to -50 degrees Celsius. So, the algorithm would classify that Mary should feel warm at -8 degrees Celsius. But this likely is not true using our common sense and knowledge. In more complex examples, we may be seduced by the results of the analysis to make false conclusions due to our lack of expertise. But remember that data science makes use of substantive and expert knowledge, not only data analysis. To make good conclusions, we should have a good understanding of the problem and our data.

    The algorithm further says that at 22 degrees Celsius, Mary should feel warm, and there is no doubt in that, as 22 degrees Celsius is higher than 20 degrees Celsius and a human being feels warmer with a higher temperature; again, a trivial use of our knowledge. For 15 degrees Celsius, the algorithm would deem Mary to feel warm, but our common sense we may not be that certain of this statement.

To be able to use our algorithm to yield better results, we should collect more data. For example, if we find out that Mary feels cold at 14 degrees Celsius, then we have a data instance that is very close to 15 degrees and, thus, we can guess with a higher certainty that Mary would feel cold at a temperature of 15 degrees.

2. The nature of the data we are dealing with is just one-dimensional and also partitioned into two parts, cold and warm, with the property: the higher the temperature, the warmer a person feels. Also, even if we know how Mary feels at temperatures, -40, -39, ..., 39, 40, we still have a very limited amount of data instances - just one around every degree Celsius. For these reasons, it is best to just look at one closest neighbor.
3. The discrepancies in the data can be caused by inaccuracy in the tests carried out. This could be mitigated by performing more experiments.

   Apart from inaccuracy, there could be other factors that influence how Mary feels: for example, the wind speed, humidity, sunshine, how warmly Mary is dressed (if she has a coat with jeans, or just shorts with a sleeveless top, or even a swimming suit), if she was wet or dry. We could add these additional dimensions (wind speed and how dressed) into the vectors of our data points. This would provide more, and better quality, data for the algorithm and, consequently, better results could be expected.

   If we have only temperature data, but more of it (for example, 10 instances of classification for every degree Celsius), then we could increase the *k* and look at more neighbors to determine the temperature more accurately. But this purely relies on the availability of the data. We could adapt the algorithm to yield the classification based on all the neighbors within a certain distance d rather than classifying based on the k-closest neighbors. This would make the algorithm work well in both cases when we have a lot of data within the close distance, but also even if we have just one data instance close to the instance that we want to classify.

4. For this purpose, one can use cross-validation (consult the *Cross-validation* section in the *Appendix A - Statistics*) to determine the value of *k* with the highest accuracy. One could separate the available data from the partial map of Italy into learning and test data, For example, 80% of the classified pixels on the map would be given to a k-NN algorithm to complete the map. Then the remaining 20% of the classified pixels from the partial map would be used to calculate the percentage of the pixels with the correct classification by the k-NN algorithm.

5. Without data rescaling, Peter's closest neighbor has an annual income of 78,000 USD and is aged 25. This neighbor does not own a house.
6. After data rescaling, Peter's closet neighbor has annual income of 60,000 USD and is aged 40. This neighbor owns a house.

7. To design a metric that accurately measures the similarity distance between the two documents, we need to select important words that will form the dimensions of the frequency vectors for the documents. The words that do not determine the semantic meaning of a documents tend to have an approximately similar frequency count across all the documents. Thus, instead, we could produce a list with the relative word frequency counts for a document. For example, we could use the following definition:

$$\text{relative\_frequency\_count}(\text{word}, \text{document}) = \frac{\text{frequency\_count}(\text{word}, \text{document})}{\text{frequency\_count}(\text{word}, \text{whole\_corpus})}$$

Then the document could be represented by an N-dimensional vector consisting of the word frequencies for the N words with the highest relative frequency count. Such a vector will tend to consist of the more important words than a vector of the N words with the highest frequency count.

2
Naive Bayes

A naive Bayes classification algorithm assigns a class to an element of a set which is most probable according to Bayes' theorem.

Let *A* and *B* be probabilistic events. *P(A)* the probability of *A* being true. *P(A|B)* the conditional probability of *A* being true given *B* is true. Then, **Bayes' theorem** states the following:

$$P(A|B)=(P(B|A) * P(A))/P(B)$$

P(A) is the prior probability of *A* being true without the knowledge of the probability of *P(B)* and *P(B|A)*. *P(A|B)* is the posterior probability of *A* being true, taking into consideration additional knowledge about the probability of *B* being true.

In this chapter, you will learn the following:

- How to apply Bayes' theorem in a basic way to compute the probability of a medical test being correct in simple example Medical test
- To grasp Bayes' theorem by proving its statement above and its extension
- How to apply Bayes' theorem differently for independent and dependent variables in examples Playing chess
- How to apply Bayes' theorem for discrete random variables in examples Medical test and Playing chess; and for continuous random variables in example Gender classification using the probability distribution of the continuous random variable
- To implement in Python an algorithm calculating the posterior probabilities using Bayes' theorem in section Implementation of naive Bayes classifier
- By verifying your understanding through solving problems in the end of the chapter to discern in what situations Bayes' theorem is an appropriate method of analysis and when it is not

Medical test - basic application of Bayes' theorem

A patient takes a special cancer test which has the accuracy *test_accuracy=99.9%*: if the result is positive, then *99.9%* of the patients tested will suffer from the special type of cancer. *99.9%* of the patients with a negative result do not suffer from the cancer.

Suppose that a patient is tested and scores positive on the test. What is the probability that a patient suffers from the special type of cancer?

Analysis:

We will use Bayes' theorem to find out the probability of the patient having the cancer:

*P(cancer|test_positive)=(P(test_positive|cancer) * P(cancer))/P(test_positive)*

To know the prior probability that a patient has the cancer, we have to find out how frequently the cancer occurs among people. Say that we find out that 1 person in 100,000 suffers from this kind of cancer. Then *P(cancer)=1/100,000*. So, *P(test_positive|cancer) = test_accuracy=99.9%=0.999* given by the accuracy of the test.

P(test_positive) has to be computed:

*P(test_positive)=P(test_positive|cancer)*P(cancer)+P(test_positive|no_cancer)*P(no_cancer)*

*= test_accuracy*P(cancer)+(1-test_accuracy)*(1-P(cancer))*

*= 2*test_accuracy*P(cancer)+1-test_accuracy-P(cancer)*

Therefore, we can compute the following:

*P(cancer|test_positive) = (test_accuracy * P(cancer))/(2 * test_accuracy * P(cancer)+1-test_accuracy-P(cancer))*

*= 0.999 * 0.00001 / (2 * 0.999 * 0.00001 + 1 - 0.999-0.00001)*

= 0.00989128497 which is approximately 1%

So, even if the result of the test is positive and the test has accuracy is *99.9%*, the probability of the patient having the tested type of cancer is only approximately 1%. This probability of having the cancer after taking the test is relatively low when compared to the high accuracy of the test, but is much higher than the probability of 1 in *100,000 (0.001%)*, as known prior to taking the test based on its occurrence in the population.

Proof of Bayes' theorem and its extension

Bayes' theorem states the following:

$$P(A|B)=[P(B|A) * P(A)]/P(B)$$

Proof:

We can prove this theorem using elementary set theory on the probability spaces of the events *A* and *B*. That is, here, a probability event will be defined as the set of the possible outcomes in the probability space:

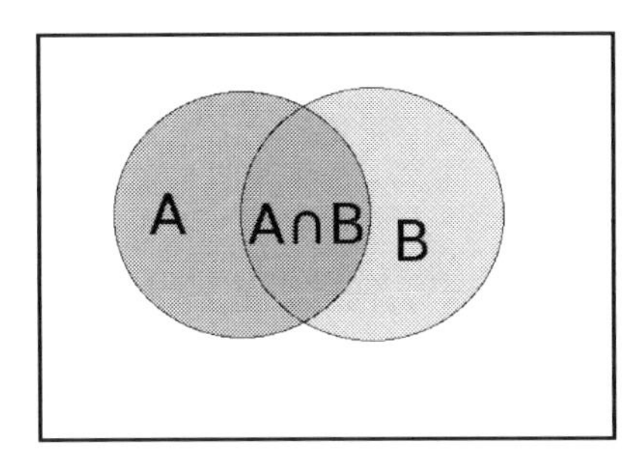

Figure 2.1: Probability space for the two events

From figure 2.1 above, we can state the following relationships:

$P(A \mid B)=P(A \cap B)/P(B)$

$P(B \mid A)=P(A \cap B)/P(A)$

Rearranging these relationships, we get the following:

$P(A \cap B)=P(A|B)*P(B)$

$P(A \cap B)=P(B|A)*P(A)$

$P(A|B)*P(B)=P(B|A)*P(A)$

This is, in fact, Bayes' theorem:

$P(A|B)=P(B|A)*P(A)/P(B)$

This concludes the proof.

Extended Bayes' theorem

We can extend Bayes' theorem taking into consideration more probability events. Suppose that the events $B_1,...,B_n$ are conditionally independent given A. Let $\sim A$ denote the complement of A. Then:

$P(A|B_1,...,B_n) = P(B_1,...,B_n|A) * P(A) / P(B_1,...,B_n)$

$= [P(B_1|A) * ... * P(B_n|A) * P(A)] / [P(B_1|A) * ... * P(B_n|A) * P(A) + P(B_1|\sim A) * ... * P(B_n|\sim A) * P(\sim A)]$

Proof:

Since the events $B_1,...,B_n$ are conditionally independent given A (and also given $\sim A$), we have the following:

$P(B_1,...,B_n|A)=P(B_1|A) * ... * P(B_n|A)$

Applying the simple form of Bayes' theorem and this fact, we thus have the following:

$P(A|B_1,...,B_n) = P(B_1,...,B_n|A) * P(A) / P(B_1,...,B_n)$

$= P(B_1|A) * ... * P(B_n|A) * P(A) / [P(B_1,...,B_n|A)*P(A)+P(B_1,...,B_n|\sim A)*P(\sim A)]$

$= [P(B_1|A) * ... * P(B_n|A) * P(A)] / [P(B_1|A) * ... * P(B_n|A) * P(A) + P(B_1|\sim A) * ... * P(B_n|\sim A) * P(\sim A)]$

This completes the proof as required.

Playing chess - independent events

Suppose we are given the following table of data with the conditions for our friend playing a game of chess with us in a park outside:

Temperature	Wind	Sunshine	Play
Cold	Strong	Cloudy	No
Warm	Strong	Cloudy	No
Warm	None	Sunny	Yes
Hot	None	Sunny	No
Hot	Breeze	Cloudy	Yes
Warm	Breeze	Sunny	Yes
Cold	Breeze	Cloudy	No
Cold	None	Sunny	Yes
Hot	Strong	Cloudy	Yes
Warm	None	Cloudy	Yes
Warm	Strong	Sunny	?

We would like to find out using Bayes' theorem whether our friend would like to play a game of chess with us in the park given that the temperature is warm, the wind is strong, and it is sunny.

Analysis:

In this case, we may want to consider temperature, wind, and sunshine as the independent random variables. The formula for the extended Bayes' theorem when adapted becomes the following:

P(Play=Yes|Temperature=Warm,Wind=Strong,Sunshine=Sunny)=R/(R+~R)

Here, *R = P(Temperature=Warm|Play=Yes)* P(Wind=Strong|Play=Yes) * P(Sunshine=Sunny|Play=Yes) * P(Play=Yes)*, and

~R = P(Temperature=Warm|Play=No) P(Wind=Strong|Play=No) * P(Sunshine=Sunny|Play=No) * P(Play=No)*.

Let us count the number of columns in the table with all known values to determine the individual probabilities.

P(Play=Yes)=6/10=3/5 since there are 10 columns with complete data and 6 of them have the value Yes for the attribute Play.

P(Temperature=Warm|Play=Yes)=3/6=1/2 since there are 6 columns with the value Yes for the attribute Play and, out of them, 3 have the value Warm for the attribute Temperature. Similarly, we have the following:

P(Wind=Strong|Play=Yes)=1/6

P(Sunshine=Sunny|Play=Yes)=3/6=1/2

P(Play=No)=4/10=2/5

P(Temperature=Warm|Play=No)=1/4

P(Wind=Strong|Play=No)=2/4=1/2

P(Sunshine=Sunny|Play=No)=1/4

Thus *R=(1/2)*(1/6)*(1/2)*(3/5)=1/40* and *~R=(1/4)*(1/2)*(1/4)*(2/5)=1/80*. Therefore, we have the following:

P(Play=Yes|Temperature=Warm,Wind=Strong,Sunshine=Sunny)= R/(R+~R)=2/3~67%

Therefore, our friend is likely to be happy to play chess with us in the park in the stated weather conditions with a probability of about 67%. Since this is a majority, we could classify the data vector (*Temperature=Warm,Wind=Strong, Sunshine=Sunny*) to be in the class *Play=Yes*.

Implementation of naive Bayes classifier

We implement a program calculating the probability of a data item belonging to a certain class using Bayes' theorem:

```
# source_code/2/naive_bayes.py
# A program that reads the CSV file with the data and returns
# the Bayesian probability for the unknown value denoted by ? to
# belong to a certain class.
# An input CSV file should be of the following format:
```

```
# 1. items in a row should be separated by a comma ','
# 2. the first row should be a heading - should contain a name for each
# column of the data.
# 3. the remaining rows should contain the data itself - rows with
# complete and rows with the incomplete data.
# A row with complete data is the row that has a non-empty and
# non-question mark value for each column. A row with incomplete data is
# the row that has the last column with the value of a question mark ?.
# Please, run this file on the example chess.csv to understand this help
# better:
# $ python naive_bayes.py chess.csv

import imp
import sys
sys.path.append('../common')
import common  # noqa

# Calculates the Baysian probability for the rows of incomplete data and
# returns them completed by the Bayesian probabilities. complete_data
# are the rows with the data that is complete and are used to calculate
# the conditional probabilities to complete the incomplete data.
def bayes_probability(heading, complete_data, incomplete_data,
                      enquired_column):
    conditional_counts = {}
    enquired_column_classes = {}
    for data_item in complete_data:
        common.dic_inc(enquired_column_classes,
                       data_item[enquired_column])
        for i in range(0, len(heading)):
            if i != enquired_column:
                common.dic_inc(
                    conditional_counts, (
                        heading[i], data_item[i],
                        data_item[enquired_column]))

    completed_items = []
    for incomplete_item in incomplete_data:
        partial_probs = {}
        complete_probs = {}
        probs_sum = 0
        for enquired_group in enquired_column_classes.items():
            # For each class in the of the enquired variable A calculate
            # the probability P(A)*P(B1|A)*P(B2|A)*...*P(Bn|A) where
            # B1,...,Bn are the remaining variables.
            probability = float(common.dic_key_count(
                enquired_column_classes,
                enquired_group[0])) / len(complete_data)
            for i in range(0, len(heading)):
```

```
                if i != enquired_column:
                    probability = probability * (float(
                        common.dic_key_count(
                            conditional_counts, (
                                heading[i], incomplete_item[i],
                                enquired_group[0]))) / (
                        common.dic_key_count(enquired_column_classes,
                                             enquired_group[0])))
            partial_probs[enquired_group[0]] = probability
            probs_sum += probability

        for enquired_group in enquired_column_classes.items():
            complete_probs[enquired_group[0]
                           ] = partial_probs[enquired_group[0]
                                             ] / probs_sum
        incomplete_item[enquired_column] = complete_probs
        completed_items.append(incomplete_item)
    return completed_items

# Program start
if len(sys.argv) < 2:
    sys.exit('Please, input as an argument the name of the CSV file.')

(heading, complete_data, incomplete_data,
 enquired_column) = common.csv_file_to_ordered_data(sys.argv[1])

# Calculate the Bayesian probability for the incomplete data
# and output it.
completed_data = bayes_probability(
    heading, complete_data, incomplete_data, enquired_column)
print completed_data

# source_code/common/common.py
# Increments integer values in a dictionary.
def dic_inc(dic, key):
    if key is None:
        pass
    if dic.get(key, None) is None:
        dic[key] = 1
    else:
        dic[key] = dic[key] + 1

def dic_key_count(dic, key):
    if key is None:
        return 0
    if dic.get(key, None) is None:
        return 0
    else:
```

```
        return int(dic[key])
```

Input:

We save the data from the table in example Playing chess in the following CSV file:

```
source_code/2/naive_bayes/chess.csv
Temperature,Wind,Sunshine,Play
Cold,Strong,Cloudy,No
Warm,Strong,Cloudy,No
Warm,None,Sunny,Yes
Hot,None,Sunny,No
Hot,Breeze,Cloudy,Yes
Warm,Breeze,Sunny,Yes
Cold,Breeze,Cloudy,No
Cold,None,Sunny,Yes
Hot,Strong,Cloudy,Yes
Warm,None,Cloudy,Yes
Warm,Strong,Sunny,?
```

Output:

We provide the file `chess.csv` as the input to the Python program calculating the probabilities of the data item (*Temperature=Warm,Wind=Strong, Sunshine=Sunny*) belonging to the classes present in the file: *Play=Yes* and *Play=No*. As we found out earlier manually, the data item belongs with a higher probability to the class *Play=Yes*. Therefore we classify the data item into that class:

```
$ python naive_bayes.py chess.csv
[
    ['Warm', 'Strong', 'Sunny', {
        'Yes': 0.6666666666666666,
        'No': 0.33333333333333337
    }]
]
```

Playing chess - dependent events

Suppose that we would like to find out again if our friend would like to play chess in the park with us in a park in Cambridge, UK. But, this time, we are given different input data:

Temperature	Wind	Season	Play
Cold	Strong	Winter	No

Warm	Strong	Autumn	No
Warm	None	Summer	Yes
Hot	None	Spring	No
Hot	Breeze	Autumn	Yes
Warm	Breeze	Spring	Yes
Cold	Breeze	Winter	No
Cold	None	Spring	Yes
Hot	Strong	Summer	Yes
Warm	None	Autumn	Yes
Warm	Strong	Spring	?

So, we wonder how the answer will change with this different data.

Analysis:

We may be tempted to use Bayesian probability to calculate the probability of our friend playing chess with us in the park. However, we should be careful, and ask whether the probability events are independent of each other.

In the previous example, where we used Bayesian probability, we were given the probability variables Temperature, Wind, and Sunshine. These are reasonably independent. Common sense tells us that a specific temperature or sunshine does not have a strong correlation to a specific wind speed. It is true that sunny weather results in higher temperatures, but sunny weather is common even when the temperatures are very low. Hence, we considered even sunshine and temperature reasonably independent as random variables and applied Bayes' theorem.

However, in this example, temperature and season are tightly related, especially in a location such as the UK, where we stated that the park we are interested in was placed. Unlike closer to the equator, temperatures in the UK vary greatly throughout the year. Winters are cold and summers are hot. Spring and fall have temperatures in between.

Therefore, we cannot apply Bayes' theorem here, as the random variables are dependent. However, we could still perform some analysis using Bayes' theorem on the partial data. By eliminating sufficient dependent variables, the remaining ones could turn out to be independent. Since temperature is a more specific variable than season, and the two variables are dependent, let us keep only the temperature variable. The remaining two variables, temperature and wind, are dependent.

Thus, we get the following data:

Temperature	Wind	Play
Cold	Strong	No
Warm	Strong	No
Warm	None	Yes
Hot	None	No
Hot	Breeze	Yes
Warm	Breeze	Yes
Cold	Breeze	No
Cold	None	Yes
Hot	Strong	Yes
Warm	None	Yes
Warm	Strong	?

We can keep the duplicate rows, as they give us greater evidence of the occurrence of the specific data row.

Input:

Saving the table we get the following CSV file:

```
# source_code/2/chess_reduced.csv
Temperature,Wind,Play
Cold,Strong,No
Warm,Strong,No
Warm,None,Yes
Hot,None,No
Hot,Breeze,Yes
Warm,Breeze,Yes
Cold,Breeze,No
Cold,None,Yes
Hot,Strong,Yes
Warm,None,Yes
Warm,Strong,?
```

Output:

We input the saved CSV file into the program `naive_bayes.py`. We get the following result:

```
python naive_bayes.py chess_reduced.csv
[['Warm', 'Strong', {'Yes': 0.49999999999999994, 'No': 0.5}]]
```

The first class, `Yes`, is going to be true with the probability 50%. The numerical difference resulted from using Python's non-exact arithmetic on the float numerical data type. The second class, No, has the same probability, 50%, of being true. We, thus, cannot make a reasonable conclusion with the data that we have about the class of the vector (`Warm`, `Strong`). However, we probably have already noticed that this vector already occurs in the table with the resulting class `No`. Hence, our guess would be that this vector should just happen to exist in one class, `No`. But, to have greater statistical confidence, we would need more data or more independent variables involved.

Gender classification - Bayes for continuous random variables

So far, we have been given a probability event that belonged to one of a finite number of classes, for example, a temperature was classified as cold, warm, or hot. But how would we calculate the posterior probability if we were given the temperature in degrees Celsius instead?

For this example, we are given five men and five women with their heights as in the following table:

Height in cm	Gender
180	Male
174	Male
184	Male
168	Male
178	Male
170	Female
164	Female

155	Female
162	Female
166	Female
172	?

Suppose that the next person has the height 172cm. What gender is that person more likely to be and with what probability?

Analysis:

One approach to solving this problem could be to assign classes to the numerical values, for example, the people with a height between 170 cm and 179 cm would be in the same class. With this approach, we may end up with a few classes that are very wide, for example, with a high cm range, or with classes that are more precise but have fewer members and so the power of Bayes cannot be manifested well. Similarly, using this method, we would not consider that the classes of height intervals in cm *[170,180)* and *[180,190)* are closer to each other than the classes *[170,180)* and *[190,200)*.

Let us remind ourselves of the Bayes' formula here:

*P(male|height)=P(height|male)*P(male)/P(height)*

*=P(height|male)*P(male)/[P(height|male)*P(male)+P(height|female)*P(female)]*

Expressing the formula in the final form above removes the need to normalize the *P(height|male)* and *P(height)* to get the correct probability of a person being male based on the measured height.

Assuming that the height of people is distributed normally, we could use a normal probability distribution to calculate *P(male|height)*. We assume *P(male)=0.5*, that is, that it is equally likely that the person to be measured is of either gender. A normal probability distribution is determined by the mean μ and the variance σ^2 of the population:

$$f(x|\mu,\sigma^2) = \frac{e^{\frac{-(x-\mu)^2}{2\sigma^2}}}{\sqrt{2\sigma^2\pi}}$$

Gender	Mean of height	Variance of height
Male	176.8	37.2
Female	163.4	30.8

Thus we could calculate the following:

P(height=172|male)=exp[-(172- 176.8)2/(2*37.2)]/[sqrt(2*37.2*π)]=0

P(height=172|female)=exp[-(172- 163.4)2/(2*30.8)]/[sqrt(2*30.8*π)]=0.02163711333

Note that these are not the probabilities, just the values of the probability density function. However, from these values, we can already observe that a person with a measured height 172 cm is more likely to be male than female because P(height=172|male)>P(height=172|female). To be more precise:

P(male|height=172)=P(height=172|male)*P(male)/[P(height=172|male)*P(male)+P(height=172|female)*P(female)]

=0.04798962999*0.5/[0.04798962999*0.5+0.02163711333*0.5]=0.68924134178~68.9%

Therefore, the person with the measured height 172 cm is a male with a probability of 68.9%.

Summary

Bayes' theorem states the following:

P(A|B)=(P(B|A) * P(A))/P(B)

Here, P(A|B) is the conditional probability of A being true given that B is true. It is used to update the value of the probability that A is true given the new observations about other probabilistic events. This theorem can be extended to a statement with multiple random variables:

P(A|B1,...,Bn)=[P(B1|A) * ... * P(Bn|A) * P(A)] / [P(B1|A) * ... * P(Bn|A) * P(A) + P(B1|~A) * ... * P(Bn|~A) * P(~A)]

The random variables *B1,...,Bn* have to be independent conditionally given *A*. The random variables can be discrete or continuous and follow some probability distribution, for example, normal (Gaussian) distribution.

For the case of a discrete random variable, it would be best to ensure you have a data item for each value of a discrete random variable given any of the conditions (value of *A*) by collecting enough data.

The more independent random variables we have, the more accurately we can determine the posterior probability. However, the greater danger there is that some of these variables could be dependent, resulting in imprecise final results. When the variables are dependent, we may eliminate some of the dependent variables and consider only mutually independent variables, or consider another algorithm as an approach to solving the data science problem.

Problems

1. A patient is tested for having a virus V. The accuracy of the test is 98%. This virus V is currently present in 4 out of 100 people in the region of the patient:

- **a)** What is the probability that a patient suffers from the virus V if they tested positive?
- **b)** What is the probability that a patient can still suffer from the disease if the result of the test was negative?

2. Apart from assessing the patients for suffering from the virus V (in question 2.1.), by using the test, a doctor usually also checks for other symptoms. According to a doctor, about 85% of patients with symptoms such as fever, nausea, abdominal discomfort, and malaise suffer from the virus V:

- **a)** What is the probability that a patient is suffering from the virus V if they have the symptoms mentioned above and their test result for the virus V is positive?
- **b)** How likely is it the patient is suffering from the virus V if they have the symptoms mentioned above, but the result of the test is negative?

3. On a certain island, 1 in 2 tsunamis are preceded by an earthquake. There have been 4 tsunamis and 6 earthquakes in the past 100 years. A seismological station recorded an earthquake in the ocean near the island. What is the probability that it will result in a tsunami?

4. Patients are tested with four independent tests on whether they have a certain illness:

Test1 positive	Test2 positive	Test3 positive	Test4 positive	Illness
Yes	Yes	Yes	No	Yes
Yes	Yes	No	Yes	Yes
No	Yes	No	No	No
Yes	No	No	No	No
No	No	No	No	No
Yes	Yes	Yes	Yes	Yes
Yes	No	Yes	Yes	Yes
No	Yes	No	No	No
No	No	Yes	Yes	No
Yes	Yes	No	Yes	Yes
Yes	No	Yes	No	Yes
Yes	No	No	Yes	Yes
No	Yes	Yes	No	?

We have taken a new patient, for whom the second and third tests are positive and the first and fourth are negative. What is the probability that they suffer from the illness?

5. We are given the following table of which words an email contains and whether it is spam or not:

Money	Free	Rich	Naughty	Secret	Spam
No	No	Yes	No	Yes	Yes
Yes	Yes	Yes	No	No	Yes
No	No	No	No	No	No
No	Yes	No	No	No	Yes
Yes	No	No	No	No	No
No	Yes	No	Yes	Yes	Yes

No	Yes	No	Yes	No	Yes
No	No	No	Yes	No	Yes
No	Yes	No	No	No	No
No	No	No	No	Yes	No
Yes	Yes	Yes	No	Yes	Yes
Yes	No	No	No	Yes	Yes
No	Yes	Yes	No	No	No
Yes	No	Yes	No	Yes	?

- **a)** What is the result of the naive Bayes algorithm when given an email that contains the words money, rich, and secret, but does not contain the words free and naughty?
- **b)** Do you agree with the result of the algorithm? Is the naive Bayes algorithm, as used here, a good method to classify email? Justify your answers.

6. Gender classification. Assume we are given the following data about 10 people:

Height in cm	**Weight in kg**	**Hair length**	**Gender**
180	75	Short	Male
174	71	Short	Male
184	83	Short	Male
168	63	Short	Male
178	70	Long	Male
170	59	Long	Female
164	53	Short	Female
155	46	Long	Female
162	52	Long	Female
166	55	Long	Female
172	60	Long	?

What is the probability that the 11th person with a height of 172cm, weight of 60kg, and long hair is a man?

Analysis:

1. Before the patient is given the test, the probability that he suffers from the virus is 4%, *P(virus)=4%=0.04*. The accuracy of the test is *test_accuracy=98%=0.98*. We apply the formula from the medical test example:

   *P(test_positive)=P(test_positive|virus)*P(virus)+P(test_positive|virus)*P(no_virus)*

   *= test_accuracy*P(virus)+(1-test_accuracy)*(1-P(virus))*

   *= 2*test_accuracy*P(virus)+1-test_accuracy-P(virus)*

   Therefore, we have the following:

- **a)** *P(virus|test_positive)=P(test_positive|virus)*P(virus)/P(test_positive)*

   *=test_accuracy*P(virus)/P(test_positive)*

   *=test_accuracy*P(virus)/[2*test_accuracy*P(virus)+1-test_accuracy-P(virus)]*

   *=0.98*0.04/[2*0.98*0.04+1-0.98-0.04]=0.67123287671~67%*

   Therefore, there is a probability of about 67% that a patient suffers from the virus V if the result of the test is positive:

- **b)** *P(virus|test_negative)=P(test_negative|virus)*P(virus)/P(test_negative)*

   *=(1-test_accuracy)*P(virus)/[1-P(test_positive)]*

   *=(1-test_accuracy)*P(virus)/[1-2*test_accuracy*P(virus)-1+test_accuracy+P(virus)]*

   *=(1-test_accuracy)*P(virus)/[test_accuracy+P(virus)-2*test_accuracy*P(virus)]*

   *=(1-0.98)*0.04/[0.98+0.04-2*0.98*0.04]=0.000849617672~0.08%*

   If the test is negative, a patient can still suffer from the virus V with a probability of 0.08%.

2. Here, we can assume that symptoms and a positive test result are conditionally independent events given that a patient suffers from virus V. The variables we have are the following:

    P(virus)=0.04

    test_accuracy=0.98

    symptoms_accuracy=85%=0.85

    Since we have two independent random variables, we will use an extended Bayes' theorem:

- **a)** *Let R=P(test_positive|virus)*P(symptoms|virus)*P(virus)*

    *=test_accuracy*symptoms_accuracy*P(virus)*

    *=0.98*0.85*0.04=0.03332*

    *~R=P(test_positive|~virus)*P(symptoms|~virus)*P(~virus)*

    =(1-test_accuracy)(1-symptoms_accuracy)*(1-P(virus))*

    =(1-0.98)(1-0.85)*(1-0.04)=0.00288*

    Then P(virus|test_positive,symptoms) = R/[R+~R]

    =0.03332/[0.03332+0.00288]=0.92044198895~92%.

    So, the patient with the symptoms for virus V and the positive test result for virus V suffers from the virus with a probability of approximately 92%.

Note that in the previous question, we learnt that a patient suffers from the disease with the probability of only about 67% given that the result of the test was positive. But after adding another independent random variable, the confidence increased to 92% even though the symptom assessment was reliable only on 85%. This implies that usually it is a very good idea to add as many independent random variables as possible to calculate the posterior probability with a higher accuracy and confidence.

- **b)** Here, the patient has the symptoms for the virus V, but the result of the test is negative. Thus we have the following:

  *R=P(test_negative|virus)*P(symptoms|virus)*P(virus)*

  *=(1-test_accuracy)*symptoms_accuracy*P(virus)*

  *=(1-0.98)*0.85*0.04=0.00068*

  *~R=P(test_negative|~virus)*P(symptoms|~virus)*P(~virus)*

  =test_accuracy(1-symptoms_accuracy)*(1-P(virus))*

  =0.98(1-0.85)*(1-0.04)=0.14112*

  Thus P(virus|test_negative,symptoms)=R/[R+~R]

  =0.00068/[0.00068+0.14112]=0.0047954866~0.48%

  Thus, a patient tested negative on the test, but with symptoms of virus V, has a probability of 0.48% of having the virus.

3. We apply the basic form of Bayes' theorem:

   *P(tsunami|earthquake)=P(earthquake|tsunami)*P(tsunami)/P(earthquake)*

   ~0.5(4/(365*100))/(6/(365*100))*

   *~0.5*4/6~1/3=33%*

   There is a chance of 33% that there will be a tsunami following the recorded earthquake.

Note that here we set P(tsunami) to be the probability of a tsunami happening on some particular day out of the days in the past 100 years. We used a day as a unit to calculate the probability P(earthquake) as well. If we changed the unit to an hour, week, month, and so on for both P(tsunami) and P(earthquake), the result would still be the same. What is important in the calculation is the ratio P(tsunami):P(earthquake)=4:6=2/3:1, that is, that a tsunami is 2/3 times more likely to happen than an earthquake.

4. We put the data into the program for calculating the posterior probability from the observations and get the following answer:

 [['No', 'Yes', 'Yes', 'No', {'Yes': 0.0, 'No': 1.0}]]

 By this calculation, a patient tested should not suffer from the illness. However, the probability of No seems quite high. It may be a good idea to get more data to get a more precise estimate of with what probability the patient is healthy.

5. The result of the algorithm is as follows:

 [['Yes', 'No', 'Yes', 'No', 'Yes', {'Yes': 0.8459918784779665, 'No': 0.15400812152203341}]]

 So, according to the naive Bayes algorithm, when applied to the data in the table, the email is spam with the probability of about 85%.

 This method may not be as good since the occurrence of certain words in a spam email is not independent. For example, spam emails containing the word money would likely try to convince that a victim of a spam could somehow get the money from the spammer and thus other words such as rich, secret, or free are more likely to be present in such an email as well. A nearest neighbor algorithm would seem to perform better at spam email classification. One could verify the actual methods using cross-validation.

6. For this problem, we will use the extended Bayes' theorem for both continuous and discrete random variables:

 P(male|height=172cm,weight=60kg,hair=long)=R/[R+~R]

 *where R=P(height=172cm|male)*P(weight=60kg|male)*P(hair=long|male)*P(male)*

 *~R=P(height=172cm|female)*P(weight=60kg|female)*P(hair=long|female)*P(female)*

Let us summarize the given information in the following tables:

Gender	Mean of height	Variance of height
Male	176.8	37.2
Female	163.4	30.8

Gender	Mean of weight	Variance of weight
Male	72.4	53.8
Female	53	22.5

From this data, let us determine other values needed to determine the final probability of the person being male:

P(height=172cm|male)=0.04798962999

*P(weight=60kg|male)=exp[-(60- 72.4)2/(2*53.8)]/[sqrt(2*53.8*π)]=0.01302907931*

P(hair=long|male)=0.2

P(male)=0.5 by assumption

P(height=172cm|female)=0.02163711333

*P(weight=60kg|female)=exp[-(60- 53)2/(2*22.5)]/[sqrt(2*22.5*π)]=0.02830872899*
P(hair=long|female)=0.8

P(female)=0.5 by assumption, Hence, we have the following:

*R=0.04798962999*0.01302907931*0.2*0.5=0.00006252606*

*~R=0.02163711333*0.02830872899*0.8*0.5=0.00024500767*

P(male|height=172cm,weight=60kg,hair=long)

=0.00006252606/[0.00006252606+0.00024500767]=0.2033144787~20.3%

Therefore, the person with height 172 cm, weight 60 kg, and long hair is a male with a probability of 20.3%. Thus, they are more likely to be female.

3
Decision Trees

A decision tree is the arrangement of the data in a tree structure where, at each node, data is separated to different branches according to the value of the attribute at the node.

To construct a decision tree, we will use a standard ID3 learning algorithm that chooses an attribute that classifies the data samples in the best possible way to maximize the information gain - a measure based on information entropy.

In this chapter, you will learn:

- What a decision tree is and how to represent data in a decision tree in example Swim preference
- In the section Information theory concepts of information entropy and information gain theoretically first, then practically applying on example Swim preference
- ID3 algorithm constructing a decision tree from the training data and its implementation in Python
- How to classify new data items using the constructed decision tree in example Swim preference
- How to provide an alternative analysis using decision trees to the problem Playing chess from the previous chapter and how the results of two different algorithms applied may differ
- Verifying your understanding at the exercise section when to use and when not to use decision trees as a method of analysis
- How to deal with data inconsistencies during decision tree construction in example Going shopping

Swim preference - representing data with decision tree

For example, we may have certain preferences on whether we would swim or not. This can be recorded in the table as follows:

Swimming suit	Water temperature	Swim preference
None	Cold	No
None	Warm	No
Small	Cold	No
Small	Warm	No
Good	Cold	No
Good	Warm	Yes

Data in this table can be represented alternatively with the following decision tree, for example:

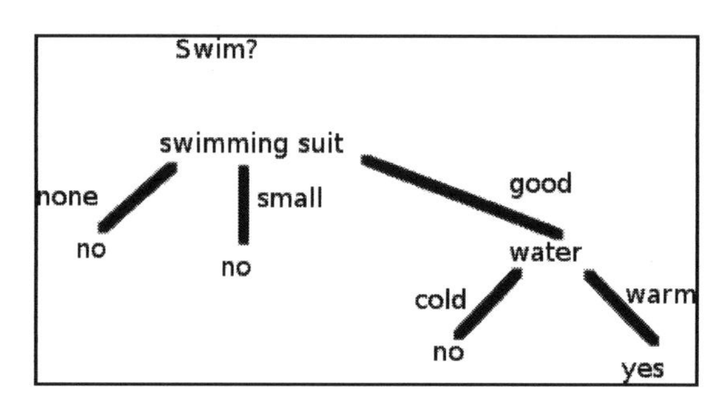

Figure 3.1.: Decision tree for the swim preference example

At the root node, we ask the question: does one have a swimming suit? The response to the question separates the available data into three groups, each with two rows. If the attribute `swimming suit = none`, then two rows have the attribute swim preference as no. Therefore, there is no need to ask a question about the temperature of the water as all the samples with the attribute `swimming suit = none` would be classified as no. This is also true for the attribute `swimming suit = small`. In the case of `swimming suit = good`, the remaining two rows can be divided into two classes: `no` and `yes`.

Without further knowledge, we would not be able to classify each row correctly. Fortunately, there is one more question that can be asked about each row which classifies each row correctly. For the row with the attribute `water=cold`, the swimming preference is no. For the row with the attribute `water=warm`, the swimming preference is yes.

To summarize, starting with the root node, we ask a question at every node and based on the answer, we move down the tree until we reach a leaf node where we find the class of the data item corresponding to those answers.

This is how we can use a ready-made decision tree to classify samples of the data. But it is also important to know how to construct a decision tree from the data.

Which attribute has a question at which node? How does this reflect on the construction of a decision tree? If we change the order of the attributes, can the resulting decision tree classify better than another tree?

Information theory

Information theory studies the quantification of information, its storage and communication. We introduce concepts of information entropy and information gain that are used to construct a decision tree using ID3 algorithm.

Information entropy

Information entropy of the given data measures the least amount of the information necessary to represent a data item from the given data. The unit of the information entropy is a familiar unit - a bit and a byte, a kilobyte, and so on. The lower the information entropy, the more regular the data is, the more pattern occurs in the data and thus less amount of the information is necessary to represent it. That is why compression tools on the computer can take large text files and compress them to a much smaller size, as words and word expressions keep reoccurring, forming a pattern.

Coin flipping

Imagine we flip an unbiased coin. We would like to know if the result is head or tail. How much information do we need to represent the result? Both words, head and tail, consist of four characters, and if we represent one character with one byte (8 bits) as it is standard in the ASCII table, then we would need four bytes or 32 bits to represent the result.

But the information entropy is the least amount of the data necessary to represent the result. We know that there are only two possible results - head or tail. If we agree to represent head with 0 and tail with 1, then one bit would be sufficient to communicate the result efficiently. Here the data is the space of the possibilities of the result of the coin throw. It is the set {head,tail} which can be represented as a set `{0,1}`. The actual result is a data item from this set. It turns out that the entropy of the set is 1. This is owing to that the probability of head and tail are both 50%.

Now imagine that the coin is biased and throws head 25% of the time and tail 75% of the time. What would be the entropy of the probability space `{0,1}` this time? We could certainly represent the result with one bit of the information. But can we do better? One bit is, of course, indivisible, but maybe we could generalize the concept of the information to indiscrete amounts.

In the previous example, we know nothing about the previous result of the coin flip unless we look at the coin. But in the example with the biased coin, we know that the result tail is more likely to happen. If we recorded n results of coin flips in a file representing heads with 0 and tails with 1, then about 75% of the bits there would have the value 1 and 25% of them would have the value 0. The size of such a file would be n bits. But since it is more regular (the pattern of 1s prevails in it), a good compression tool should be able to compress it to less than n bits.

To learn the theoretical bound to the compression and the amount of the information necessary to represent a data item, we define information entropy precisely.

Definition of information entropy

Suppose that we are given a probability space S with the elements $1, 2, ..., n$. The probability an element i would be chosen from the probability space is p_i. Then the information entropy of the probability space is defined as:

$E(S)=-p_1 *log_2(p_1) - ... - p_n *log_2(p_n)$ where log_2 is a binary logarithm.

So the information entropy of the probability space of unbiased coin throws is:

$E = -0.5 * log_2(0.5) - 0.5*log_2(0.5)=0.5+0.5=1$

When the coin is based with 25% chance of a head and 75% change of a tail, then the information entropy of such space is:

$E = -0.25 * log_2(0.25) - 0.75*log_2(0.75) = 0.81127812445$

which is less than 1. Thus, for example, if we had a large file with about 25% of 0 bits and 75% of 1 bits, a good compression tool should be able to compress it down to about 81.12% of its size.

Information gain

The information gain is the amount of the information entropy gained as a result of a certain procedure. For example, if we would like to know the results of three fair coins, then its information entropy is 3. But if we could look at the third coin, then the information entropy of the result for the remaining two coins would be 2. Thus, by looking at the third coin, we gained one bit information, so the information gain is 1.

We may also gain the information entropy by dividing the whole set S into sets, grouping them by a similar pattern. If we group elements by their value of an attribute A, then we define the information gain as:

$$IG(S, A) = E(S) - \sum_{v \in \text{values}(A)} \left[\frac{|S_v|}{|S|} * E(S_v) \right]$$

where S_v is a set with the elements of S that have the value v for the attribute A.

Swim preference - information gain calculation

Let us calculate the information gain for the six rows in the swim preference example by taking swimming suit as an attribute. Because we are interested whether a given row of data is classified as no or yes for the question whether one should swim, we will use the swim preference to calculate the entropy and information gain. We partition the set S by the attribute swimming suit:

$S_{none}=\{(none,cold,no),(none,warm,no)\}$

S_{small}=*{(small,cold,no),(small,warm,no)}*

S_{good}=*{(good,cold,no),(good,warm,yes)}*

The information entropy of *S* is $E(S)=-(1/6)*log_2(1/6)-(5/6)*log_2(5/6)\sim0.65002242164$.

The information entropy of the partitions is:

$E(S_{none})=-(2/2)*log_2(2/2)=-log_2(1)=0$ since all instances have the class no.

$E(S_{small})=0$ for a similar reason.

$E(S_{good})=-(1/2)*log_2(1/2)=1$

Therefore, the information gain is:

$IG(S,swimming\ suit)=E(S)-[(2/6)*E(S_{none})+(2/6)*E(S_{small})+(2/6)*E(S_{good})]$

$=0.65002242164-(1/3)=0.3166890883$

If we chose the attribute water temperature to partition the set *S*, what would be the information gain IG(S,water temperature)? The water temperature partitions the set *S* into the following sets:

S_{cold}=*{(none,cold,no),(small,cold,no),(good,cold,no)}*

S_{warm}=*{(none,warm,no),(small,warm,no),(good,warm,yes)}*

Their entropies are:

$E(S_{cold})=0$ as all instances are classified as no.

$E(S_{warm})=-(2/3)*log_2(2/3)-(1/3)*log_2(1/3)\sim0.91829583405$

Therefore, the information gain from partitioning the set S by the attribute water temperature is:

$IG(S,water\ temperature)=E(S)-[(1/2)*E(S_{cold})+(1/2)*E(S_{warm})]$

$= 0.65002242164-0.5*0.91829583405=0.19087450461$

This is less than *IG(S,swimming suit)*. Therefore, we can gain more information about the set *S* (the classification of its instances) by partitioning it per the attribute swimming suit instead of the attribute water temperature. This finding will be the basis of the ID3 algorithm constructing a decision tree in the next section.

ID3 algorithm - decision tree construction

The ID3 algorithm constructs a decision tree from the data based on the information gain. In the beginning, we start with the set S. The data items in the set S have various properties according to which we can partition the set S. If an attribute A has the values $\{v_1, ..., v_n\}$, then we partition the set S into the sets $S_1, ..., S_n$, where the set S_i is a subset of the set S, where the elements have the value v_i for the attribute A.

If each element in the set S has attributes $A_1, ..., A_m$, then we can partition the set S according to any of the possible attributes. The ID3 algorithm partitions the set S according to the attribute that yields the highest information gain. Now suppose that it was attribute A_1. Then for the set S we have the partitions $S_1, ..., S_n$, where A_1 has the possible values $\{v_1,..., v_n\}$.

Since we have not constructed any tree yet, we first place a root node into the tree. For every partition of S, we place a new branch from the root. Every branch represents one value of the selected attributes. A branch has data samples with the same value for that attribute. For every new branch, we can define a new node that will have data samples from its ancestor branch.

Once we have defined a new node, we choose another of the remaining attributes with the highest information gain for the data at that node to partition the data at that node further, then define new branches and nodes. This process can be repeated until we run out of all the attributes for the nodes or even earlier until all the data at the node has the same class of our interest. In the case of the swim preference example, there are only two possible classes for the swimming preference: class no and class yes. The last node is called a leaf node and decides the class of a data item from the data.

Swim preference - decision tree construction by ID3 algorithm

Here we describe, step by step, how an ID3 algorithm would construct a decision tree from the given data samples in the swim preference example. The initial set consists of six data samples:

```
S={(none,cold,no),(small,cold,no),(good,cold,no),(none,warm,no),(small,warm
,no),(good,warm,yes)}
```

In the previous sections, we calculated the information gains for both and the only non-classifying attributes, swimming suit and water temperature:

```
IG(S,swimming suit)=0.3166890883
IG(S,water temperature)=0.19087450461
```

Hence, we would choose the attribute swimming suit as it has a higher information gain. There is no tree drawn yet, so we start from the root node. As the attribute swimming suit has three possible values {none, small, good}, we draw three possible branches out of it for each. Each branch will have one partition from the partitioned set S: S_{none}, S_{small}, and S_{good}. We add nodes to the ends of the branches. S_{none} data samples have the same class swimming preference = no, so we do not need to branch that node by a further attribute and partition the set. Thus, the node with the data S_{none} is already a leaf node. The same is true for the node with the data S_{small}.

But the node with the data S_{good} has two possible classes for swimming preference. Therefore, we will branch the node further. There is only one non-classifying attribute left - water temperature. So there is no need to calculate the information gain for that attribute with the data S_{good}. From the node S_{good}, we will have two branches, each with the partition from the set S_{good}. One branch will have the set of the data sample $S_{good,\ cold}$*={(good,cold,no)}*, the other branch will have the partition $S_{good,\ warm}$*={(good,warm,yes)}*. Each of these two branches will end with a node. Each node will be a leaf node because each node has the data samples of the same value for the classifying attribute swimming preference.

The resulting decision tree has four leaf nodes and is the tree in the figure 3.1. - Decision tree for the swim preference example.

Implementation

We implement ID3 algorithm that constructs a decision tree for the data given in a csv file. All sources are in the chapter directory. The most import parts of the source code are given here:

```
# source_code/3/construct_decision_tree.py
# Constructs a decision tree from data specified in a CSV file.
# Format of a CSV file:
# Each data item is written on one line, with its variables separated
# by a comma. The last variable is used as a decision variable to
# branch a node and construct the decision tree.

import math
# anytree module is used to visualize the decision tree constructed by
# this ID3 algorithm.
```

```
from anytree import Node, RenderTree
import sys
sys.path.append('../common')
import common
import decision_tree

# Program start
csv_file_name = sys.argv[1]
verbose = int(sys.argv[2])  # verbosity level, 0 - only decision tree

# Define the equired column to be the last one.
# I.e. a column defining the decision variable.
(heading, complete_data, incomplete_data,
 enquired_column) = common.csv_file_to_ordered_data(csv_file_name)

tree = decision_tree.constuct_decision_tree(
    verbose, heading, complete_data, enquired_column)
decision_tree.display_tree(tree)

# source_code/common/decision_tree.py
# ***Decision Tree library ***
# Used to construct a decision tree and a random forest.
import math
import random
import common
from anytree import Node, RenderTree
from common import printfv

# Node for the construction of a decision tree.
class TreeNode:

    def __init__(self, var=None, val=None):
        self.children = []
        self.var = var
        self.val = val

    def add_child(self, child):
        self.children.append(child)

    def get_children(self):
        return self.children

    def get_var(self):
        return self.var

    def get_val(self):
        return self.val
```

```
    def is_root(self):
        return self.var is None and self.val is None

    def is_leaf(self):
        return len(self.children) == 0

    def name(self):
        if self.is_root():
            return "[root]"
        return "[" + self.var + "=" + self.val + "]"

# Constructs a decision tree where heading is the heading of the table
# with the data, i.e. the names of the attributes.
# complete_data are data samples with a known value for every attribute.
# enquired_column is the index of the column (starting from zero) which
# holds the classifying attribute.
def constuct_decision_tree(verbose, heading, complete_data,
enquired_column):
    return construct_general_tree(verbose, heading, complete_data,
                                  enquired_column, len(heading))

# m is the number of the classifying variables that should be at most
# considered at each node. m needed only for a random forest.
def construct_general_tree(verbose, heading, complete_data,
                           enquired_column, m):
    available_columns = []
    for col in range(0, len(heading)):
        if col != enquired_column:
            available_columns.append(col)
    tree = TreeNode()
    printfv(2, verbose, "We start the construction with the root node" +
                        " to create the first node of the tree.\n")
    add_children_to_node(verbose, tree, heading, complete_data,
                         available_columns, enquired_column, m)
    return tree

# Splits the data samples into the groups with each having a different
# value for the attribute at the column col.
def split_data_by_col(data, col):
    data_groups = {}
    for data_item in data:
        if data_groups.get(data_item[col]) is None:
            data_groups[data_item[col]] = []
        data_groups[data_item[col]].append(data_item)
    return data_groups

# Adds a leaf node to node.
def add_leaf(verbose, node, heading, complete_data, enquired_column):
```

```
    leaf_node = TreeNode(heading[enquired_column],
                         complete_data[0][enquired_column])
    printfv(2, verbose,
            "We add the leaf node " + leaf_node.name() + ".\n")
    node.add_child(leaf_node)

# Adds all the descendants to the node.
def add_children_to_node(verbose, node, heading, complete_data,
                         available_columns, enquired_column, m):
    if len(available_columns) == 0:
        printfv(2, verbose, "We do not have any available variables " +
                "on which we could split the node further, therefore " +
                "we add a leaf node to the current branch of the tree. ")
        add_leaf(verbose, node, heading, complete_data, enquired_column)
        return -1

    printfv(2, verbose, "We would like to add children to the node " +
            node.name() + ".\n")

    selected_col = select_col(
        verbose, heading, complete_data, available_columns,
        enquired_column, m)
    for i in range(0, len(available_columns)):
        if available_columns[i] == selected_col:
            available_columns.pop(i)
            break

    data_groups = split_data_by_col(complete_data, selected_col)
    if (len(data_groups.items()) == 1):
        printfv(2, verbose, "For the chosen variable " +
                heading[selected_col] +
                " all the remaining features have the same value " +
                complete_data[0][selected_col] + ". " +
                "Thus we close the branch with a leaf node. ")
        add_leaf(verbose, node, heading, complete_data, enquired_column)
        return -1

    if verbose >= 2:
        printfv(2, verbose, "Using the variable " +
                heading[selected_col] +
                " we partition the data in the current node, where" +
                " each partition of the data will be for one of the " +
                "new branches from the current node " + node.name() +
                ". " + "We have the following partitions:\n")
        for child_group, child_data in data_groups.items():
            printfv(2, verbose, "Partition for " +
                    str(heading[selected_col]) + "=" +
                    str(child_data[0][selected_col]) + ": " +
```

```
                        str(child_data) + "\n")
            printfv(
                2, verbose, "Now, given the partitions, let us form the " +
                            "branches and the child nodes.\n")
        for child_group, child_data in data_groups.items():
            child = TreeNode(heading[selected_col], child_group)
            printfv(2, verbose, "\nWe add a child node " + child.name() +
                    " to the node " + node.name() + ". " +
                    "This branch classifies %d feature(s): " +
                    str(child_data) + "\n", len(child_data))
            add_children_to_node(verbose, child, heading, child_data, list(
                available_columns), enquired_column, m)
            node.add_child(child)
        printfv(2, verbose,
                "\nNow, we have added all the children nodes for the " +
                "node " + node.name() + ".\n")

    # Selects an available column/attribute with the highest
    # information gain.
    def select_col(verbose, heading, complete_data, available_columns,
                   enquired_column, m):
        # Consider only a subset of the available columns of size m.
        printfv(2, verbose,
                "The available variables that we have still left are " +
                str(numbers_to_strings(available_columns, heading)) + ". ")
        if len(available_columns) < m:
            printfv(
                2, verbose, "As there are fewer of them than the " +
                            "parameter m=%d, we consider all of them. ", m)
            sample_columns = available_columns
        else:
            sample_columns = random.sample(available_columns, m)
            printfv(2, verbose,
                    "We choose a subset of them of size m to be " +
                    str(numbers_to_strings(available_columns, heading)) +
                    ".")

        selected_col = -1
        selected_col_information_gain = -1
        for col in sample_columns:
            current_information_gain = col_information_gain(
                complete_data, col, enquired_column)
            # print len(complete_data),col,current_information_gain
            if current_information_gain > selected_col_information_gain:
                selected_col = col
                selected_col_information_gain = current_information_gain
        printfv(2, verbose,
                "Out of these variables, the variable with " +
```

```
            "the highest information gain is the variable " +
            heading[selected_col] +
            ". Thus we will branch the node further on this " +
            "variable. " +
            "We also remove this variable from the list of the " +
            "available variables for the children of the current node. ")
    return selected_col

# Calculates the information gain when partitioning complete_data
# according to the attribute at the column col and classifying by the
# attribute at enquired_column.
def col_information_gain(complete_data, col, enquired_column):
    data_groups = split_data_by_col(complete_data, col)
    information_gain = entropy(complete_data, enquired_column)
    for _, data_group in data_groups.items():
        information_gain -= (float(len(data_group)) / len(complete_data)
                             ) * entropy(data_group, enquired_column)
    return information_gain

# Calculates the entropy of the data classified by the attribute
# at the enquired_column.
def entropy(data, enquired_column):
    value_counts = {}
    for data_item in data:
        if value_counts.get(data_item[enquired_column]) is None:
            value_counts[data_item[enquired_column]] = 0
        value_counts[data_item[enquired_column]] += 1
    entropy = 0
    for _, count in value_counts.items():
        probability = float(count) / len(data)
        entropy -= probability * math.log(probability, 2)
    return entropy
```

Program input:

We input the data from the swim preference example into the program to construct a decision tree:

```
# source_code/3/swim.csv
swimming_suit,water_temperature,swim
None,Cold,No
None,Warm,No
Small,Cold,No
Small,Warm,No
Good,Cold,No
Good,Warm,Yes
```

Program output:

We construct a decision tree from the data file `swim.csv` with the verbosity set to 0. The reader is encouraged to set the verbosity to `2` to see a detailed explanation how exactly the decision tree is constructed:

```
$ python construct_decision_tree.py swim.csv 0
Root
├── [swimming_suit=Small]
│ ├── [water_temperature=Cold]
│ │ └── [swim=No]
│ └── [water_temperature=Warm]
│ └── [swim=No]
├── [swimming_suit=None]
│ ├── [water_temperature=Cold]
│ │ └── [swim=No]
│ └── [water_temperature=Warm]
│ └── [swim=No]
└── [swimming_suit=Good]
    ├── [water_temperature=Cold]
    │ └── [swim=No]
    └── [water_temperature=Warm]
        └── [swim=Yes]
```

Classifying with a decision tree

Once we have constructed a decision tree from the data with the attributes $A_1, ..., A_m$ and the classes $\{c_1, ..., c_k\}$, we can use this decision tree to classify a new data item with the attributes $A_1, ..., A_m$ into one of the classes $\{c_1, ..., c_k\}$.

Given a new data item that we would like to classify, we can think of each node including the root as a question for data sample: What value does that data sample for the selected attribute A_i have? Then based on the answer, we select the branch of a decision tree and move further to the next node. Then another question is answered about the data sample and another until the data sample reaches the leaf node. A leaf node has an associated one of the classes $\{c_1, ..., c_k\}$ with it; for example, c_i. Then the decision tree algorithm would classify the data sample into the class c_i.

Classifying a data sample with the swimming preference decision tree

Let us construct a decision tree for the swimming preference example with the ID3 algorithm. Consider a data sample *(good, cold,?)* and we would like to use the constructed decision tree to decide into which class it should belong.

Start with a data sample at the root of the tree. The first attribute that branches from the root is swimming suit, so we ask for the value for the attribute swimming suit of the sample *(good, cold,?)*. We learn that the value of the attribute is swimming suit=good; therefore, move down the rightmost branch with that value for its data samples. We arrive at the node with the attribute water temperature and ask the question: what is the value of the attribute water temperature for the data sample *(good, cold,?)*? We learn that for that data sample, we have water temperature=cold; therefore, we move down the left branch into the leaf node. This leaf is associated with the class swimming preference=no. Therefore, the decision tree would classify the data sample *(good, cold,?)* to be in that class swimming preference; that is, to complete it to the data sample *(good, cold, no)*.

Therefore, the decision tree says that if one has a good swimming suit, but the water temperature is cold, then one would still not want to swim based on the data collected in the table.

Playing chess - analysis with decision tree

Let us take an example from the `Chapter 2`, *Naive Bayes* again:

Temperature	Wind	Sunshine	Play
Cold	Strong	Cloudy	No
Cold	Strong	Cloudy	No
Warm	None	Sunny	Yes
Hot	None	Sunny	No
Hot	Breeze	Cloudy	Yes
Warm	Breeze	Sunny	Yes
Cold	Breeze	Cloudy	No
Cold	None	Sunny	Yes

Hot	Strong	Cloudy	Yes
Warm	None	Cloudy	Yes
Warm	Strong	Sunny	?

We would like to find out if our friend would like to play chess with us outside in the park. But this time, we would like to use decision trees to find the answer.

Analysis:

We have the initial set *S* of the data samples as:

```
S={(Cold,Strong,Cloudy,No),(Warm,Strong,Cloudy,No),(Warm,None,Sunny,Yes),
(Hot,None,Sunny,No),(Hot,Breeze,Cloudy,Yes),(Warm,Breeze,Sunny,Yes),(Cold,B
reeze,Cloudy,No),(Cold,None,Sunny,Yes),(Hot,Strong,Cloudy,Yes),(Warm,None,C
loudy,Yes)}
```

First we determine the information gain for each of the three non-classifying attributes: temperature, wind, and sunshine. Possible values for temperature are cold, warm, and hot. Therefore, we will partition the set *S* into the three sets:

```
Scold={(Cold,Strong,Cloudy,No),(Cold,Breeze,Cloudy,No),(Cold,None,Sunny,Yes)}
Swarm={(Warm,Strong,Cloudy,No),(Warm,None,Sunny,Yes),(Warm,Breeze,Sunny,Yes),
(Warm,None,Cloudy,Yes)}
Shot={(Hot,None,Sunny,No),(Hot,Breeze,Cloudy,Yes),(Hot,Strong,Cloudy,Yes)}
```

We calculate the information entropies for the sets first:

$E(S)=-(4/10)*\log_2(4/10)-(6/10)*\log_2(6/10)=0.97095059445$

$E(S_{cold})=-(2/3)*\log_2(2/3)-(1/3)*\log_2(1/3)=0.91829583405$

$E(S_{warm})=-(1/4)*\log_2(1/4)-(3/4)*\log_2(3/4)=0.81127812445$

$E(S_{hot})=-(1/3)*\log_2(1/3)-(2/3)*\log_2(2/3)=0.91829583405$

Thus, $IG(S,temperature)=E(S)-[(|S_{cold}|/|S|)*E(S_{cold})+(|S_{warm}|/|S|)*E(S_{warm})+(|S_{hot}|/|S|)*E(S_{hot})]$

$=0.97095059445-[(3/10)*0.91829583405+(4/10)*0.81127812445+(3/10)*0.91829583405]$

$=0.09546184424$

Possible values for the attribute wind are none, breeze, strong. Thus we will partition the set S into the three partitions:

S_{none}={(Warm,None,Sunny,Yes),(Hot,None,Sunny,No),(Cold,None,Sunny,Yes),(Warm,None, Cloudy,Yes)}

S_{breeze}={(Hot,Breeze,Cloudy,Yes),(Warm,Breeze,Sunny,Yes),(Cold,Breeze,Cloudy,No)}

S_{strong}={(Cold,Strong,Cloudy,No),(Warm,Strong,Cloudy,No),(Hot,Strong,Cloudy,Yes)}

The information entropies of the sets are:

$E(S_{none})=0.81127812445$

$E(S_{breeze})=0.91829583405$

$E(S_{strong})=0.91829583405$

Thus, $IG(S,wind)=E(S)-[(|S_{none}|/|S|)*E(S_{none})+(|S_{breeze}|/|S|)*E(S_{breeze})+(|S_{strong}|/|S|)*E(S_{strong})]$

$= 0.97095059445-[(4/10)*0.81127812445+(3/10)*0.91829583405+(3/10)*0.91829583405]$

$= 0.09546184424$

Finally, the third attribute sunshine has two possible values, cloudy and sunny; thus, it partitions the set S into two sets:

S_{cloudy}={(Cold,Strong,Cloudy,No),(Warm,Strong,Cloudy,No),(Hot,Breeze,Cloudy,Yes), (Cold,Breeze,Cloudy,No),(Hot,Strong,Cloudy,Yes),(Warm,None,Cloudy,Yes)}

S_{sunny}={(Warm,None,Sunny,Yes),(Hot,None,Sunny,No),(Warm,Breeze,Sunny,Yes), (Cold,None,Sunny,Yes)}

The entropies of the sets are:

$E(S_{cloudy})=1$

$E(S_{sunny})=0.81127812445$

Thus, $IG(S,sunshine)=E(S)-[(|S_{cloudy}|/|S|)*E(S_{cloudy})+(|S_{sunny}|/|S|)*E(S_{sunny})]$

$=0.97095059445-[(6/10)*1+(4/10)*0.81127812445]=0.04643934467$

IG(S,wind) and *IG(S,temperature)* are greater than *IG(S,sunshine)*. Both of them are equal; therefore, we can choose any of the attribute to form the three branches; for example, the first one, temperature. In that case, each of the three branches would have data samples S_{cold}, S_{warm}, S_{hot}. At those branches, we could apply the algorithm further to form the rest of the decision tree. Instead, we will use the program to complete the tree.

Input:

```
source_code/3/chess.csv
Temperature,Wind,Sunshine,Play
Cold,Strong,Cloudy,No
Warm,Strong,Cloudy,No
Warm,None,Sunny,Yes
Hot,None,Sunny,No
Hot,Breeze,Cloudy,Yes
Warm,Breeze,Sunny,Yes
Cold,Breeze,Cloudy,No
Cold,None,Sunny,Yes
Hot,Strong,Cloudy,Yes
Warm,None,Cloudy,Yes
```

Output:

```
$ python construct_decision_tree.py chess.csv 0
Root
├── [Temperature=Cold]
│   ├── [Wind=Breeze]
│   │   └── [Play=No]
│   ├── [Wind=Strong]
│   │   └── [Play=No]
│   └── [Wind=None]
│       └── [Play=Yes]
├── [Temperature=Warm]
│   ├── [Wind=Breeze]
│   │   └── [Play=Yes]
│   ├── [Wind=None]
│   │   ├── [Sunshine=Sunny]
│   │   │   └── [Play=Yes]
│   │   └── [Sunshine=Cloudy]
│   │       └── [Play=Yes]
│   └── [Wind=Strong]
│       └── [Play=No]
└── [Temperature=Hot]
    ├── [Wind=Strong]
    │   └── [Play=Yes]
    ├── [Wind=None]
    │   └── [Play=No]
```

```
    └── [Wind=Breeze]
        └── [Play=Yes]
```

Classification:

Now that we have constructed the decision tree, we would like to use it to classify a data sample *(warm,strong,sunny,?)* into one of the two classes in the set *{no,yes}*.

We start at the root. What value does the attribute temperature have in that instance? Warm, so we go to the middle branch. What value does the attribute wind have in that instance? Strong, so the instance would fall into the class no since we have arrived already in the leaf node.

So, our friend would not want to play chess with us in the park according to the decision tree classification algorithm. Please note that the Naive Bayes' algorithm stated otherwise. An understanding of the problem is required to choose the best possible method. At other times, a method with a greater accuracy is the one that takes into consideration results of several algorithms or several classifiers, as in the case of random forest algorithm in the next chapter.

Going shopping - dealing with data inconsistency

We have the following data about the shopping preferences of our friend, Jane:

Temperature	Rain	Shopping
Cold	None	Yes
Warm	None	No
Cold	Strong	Yes
Cold	None	No
Warm	Strong	No
Warm	None	Yes
Cold	None	?

We would like to find out, using the decision trees, whether Jane would go shopping if the outside temperature was cold with no rain.

Analysis:

Here we should be careful, as there are instances of the data that have the same values for the same attributes, but have different classes; that is, `(cold,none,yes)` and `(cold,none,no)`. The program we made would form the following decision tree:

```
Root
├── [Temperature=Cold]
│    ├──[Rain=None]
│    │    └──[Shopping=Yes]
│    └──[Rain=Strong]
│    └──[Shopping=Yes]
└── [Temperature=Warm]
├──[Rain=None]
│    └──[Shopping=No]
└── [Rain=Strong]
└── [Shopping=No]
```

But at the leaf node `[Rain=None]` with the parent `[Temperature=Cold]`, there are two data samples with both classes no and yes. We cannot therefore classify an instance `(cold,none,?)` accurately. For the decision tree algorithm to work better, we would have to either provide a class at the leaf node with the greatest weight - that is, the majority class. Even better would be to collect values for more attributes for the data samples so that we can make a decision more accurately.

Therefore, in the presence of the given data, we are uncertain whether Jane would go shopping or not.

Summary

A decision tree ID3 algorithm first constructs a decision tree from the input data and then classifies a new data instance using this constructed tree. A decision tree is constructed by selecting the attribute for branching with the highest information gain. The information gain measures how much information can be learned in terms of the gain in the information entropy.

The decision tree algorithm can achieve a different result from other algorithms such as Naive Bayes' algorithm. In the next chapter, we will learn how to combine various algorithms or classifiers into a decision forest (called **random forest**) in order to achieve a more accurate result.

Problems

1. What is the information entropy of the following multisets?
   a) {1,2}, b) {1,2,3}, c) {1,2,3,4}, d) {1,1,2,2}, e) {1,1,2,3}
2. What is the information entropy of the probability space induced by the biased coin that shows heads with the probability 10% and tails with the probability 90%?
3. Let us take another example of playing chess from `Chapter 2`, *Naive Bayes*:

- **a)** What is the information gain for each of the non-classifying attributes in the table?
- **b)** What is the decision tree constructed from the given table?
- **c)** How would you classify a data sample `(warm,strong,spring,?)` according to the constructed decision tree?

Temperature	**Wind**	**Season**	**Play**
Cold	Strong	Winter	No
Warm	Strong	Autumn	No
Warm	None	Summer	Yes
Hot	None	Spring	No
Hot	Breeze	Autumn	Yes
Warm	Breeze	Spring	Yes
Cold	Breeze	Winter	No
Cold	None	Spring	Yes
Hot	Strong	Summer	Yes
Warm	None	Autumn	Yes
Warm	Strong	Spring	?

4. **Mary and temperature preferences.** Let us take the example from the `Chapter 1`, *Classification Using K Nearest Neighbors,* about the temperature preferences of Mary.

Temperature in degrees Celsius	Wind speed in kmph	Mary's perception
10	0	Cold
25	0	Warm
15	5	Cold
20	3	Warm
18	7	Cold
20	10	Cold
22	5	Warm
24	6	Warm

We would like to use decision trees to decide if our friend Mary would feel warm or cold in the room with the temperature 16 degrees Celsius with the fan of the wind speed 3km/h.

Can you please explain how a decision tree algorithm could be used here and how good it would be to use it for this example?

Analysis:

1. Here are entropy of the multisets:

*E({1,2})=-(1/2)*log2(1/2)-(1/2)*log2(1/2)=1*

*E({1,2,3})=-(1/3)*log2(1/3)-(1/3)*log2(1/3)-(1/3)*log2(1/3)=1.5849625*

*E({1,2,3,4})=-(1/4)*log2(1/4)-(1/4)*log2(1/4)-(1/4)*log2(1/4)-(1/4)*log2(1/4)=2*

*E({1,1,2,2})=-(2/4)*log2(2/4)-(2/4)*log2(2/4)=1*

*E({1,1,2,3})=-(2/4)*log2(2/4)-(1/4)*log2(1/4)-(1/4)*log2(1/4)=1.5*

Here note that the information entropy of the multisets that have more than two classes is greater than 1, so we need more than one bit of information to represent the result. But is this true for every multiset that has more than two classes of elements?

2. E({10% of heads, 90% of tails})=-0.1*$\log_2$(0.1)-(0.9)*$\log_2$(0.9)=0.46899559358
3. The information gains for the three attributes are as follows:

   IG(S,temperature)=0.0954618442383

   IG(S,wind)=0.0954618442383

   IG(S,season)=0.419973094022

   Therefore, we would choose the attribute season to branch from the root node as it has the highest information gain. Alternatively, we can put all the input data into the program to construct a decision tree:

```
Root
├── [Season=Autumn]
│    ├──[Wind=Breeze]
│    │    └──[Play=Yes]
│    ├──[Wind=Strong]
│    │    └──[Play=No]
│    └──[Wind=None]
│    └──[Play=Yes]
├── [Season=Summer]
│    ├──[Temperature=Hot]
│    │    └──[Play=Yes]
│    └──[Temperature=Warm]
│    └──[Play=Yes]
├── [Season=Winter]
│    └──[Play=No]
└── [Season=Spring]
├── [Temperature=Hot]
│    └──[Play=No]
├── [Temperature=Warm]
│    └──[Play=Yes]
└── [Temperature=Cold]
└── [Play=Yes]
```

   According to the constructed decision tree, we would classify the data sample (warm,strong,spring,?) to the class Play=Yes by going to the bottommost branch from the root node and then arriving to the leaf node by taking the middle branch.

4. Here the decision tree algorithm may not perform that well without any processing of the data. If we considered every class of a temperature, then 25 degrees Celsius would still not occur in the decision tree as it is not in the input data, so we would not be able to classify how Mary would feel at 16 degrees Celsius and at 3km/h windspeed.

   We could alternatively divide the temperature and wind speed into the intervals in order to reduce the classes, so that the resulting decision tree could classify the input instance. But it is this division, the classification of in what intervals 25 degrees Celsius and 3km/h should be, that is the fundamental part of the analysis procedure for this type of problem. Thus decision trees without any serious modification could not analyze the problem well.

4
Random Forest

A random forest is a set of random decision trees (similar to the ones described in the previous chapter), each generated on a random subset of the data. A random forest classifies the feature to belong to the class that is voted for by the majority of the random decision trees. A random forest tends to provide a more accurate classification of a feature than a decision tree because of the decreased bias and variance.

In this chapter, you will learn:

- Tree bagging (or bootstrap aggregation) technique as part of random forest construction, but that can be extended also to other algorithms and methods in data science to reduce the bias and variance and hence to improve the accuracy
- In example Swim preference to construct a random forest and classify a data item using the constructed random forest
- How to implement an algorithm in Python that would construct a random forest
- In example Playing chess the differences in the analysis of a problem by algorithms naive Bayes, decision trees and random forest
- In example Going shopping how random forest algorithm can overcome the shortcomings of decision tree algorithm and thus outperform it
- In example Going shopping how a random forest can express the level of the confidence in its classification of the feature
- In exercises how decreasing the variance of a classifier can yield more accurate results

Overview of random forest algorithm

General considerations, to begin with, we choose the number of the trees that we are going to construct for a decision forest. A random forest does not tend to overfit (unless the data is very noisy), so choosing many decision trees will not decrease the accuracy of the prediction. However, the more decision trees, the more computational power is required. Also, increasing the number of the decision trees in the forest dramatically, does not increase the accuracy of the classification much. It is important that we have sufficiently many decision trees so that most of the data is used for the classification when chosen randomly for the construction of a decision tree.

In practice, one can run the algorithm on a specific number of decision trees, increase their number, and compare the results of the classification of a smaller and a bigger forest. If the results are very similar, then there is no reason to increase the number of trees.

To simplify demonstration, in this book, we will use a small number of decision trees in a random forest.

Overview of random forest construction

We will describe how each tree is constructed in a random fashion. Given N training features, for the construction of each decision tree, we provide the data by selecting N features randomly with replacement from the initial data for the random forest. This process of selecting the data randomly with replacement for each tree is called **bootstrap aggregating** or **tree bagging**. The purpose of bootstrap aggregating is to reduce the variance and bias in the results of the classification.

Say a feature has M variables that are used to classify the feature using the decision tree. When we must make a branching decision at a node, in the ID3 algorithm we choose the variable that resulted in the highest information gain. Here in a random decision tree at each node, we consider only at most m (which is at most *M*) variables (we do not consider the ones that were already chosen) sampled in a random fashion without the replacement from the given M variables. Then out of these m variables, we choose the one that results in the highest information gain.

The rest of the construction of a random decision tree is carried out just as it was for a decision tree in the previous chapter.

Swim preference - analysis with random forest

We will use the example from the previous chapter about the swim preference. We have the same data table:

Swimming suit	Water temperature	Swim preference
None	Cold	No
None	Warm	No
Small	Cold	No
Small	Warm	No
Good	Cold	No
Good	Warm	Yes

We would like to construct a random forest from this data and use it to classify an item (Good,Cold,?).

Analysis:

We are given *M=3* variables according to which a feature can be classified. In a random forest algorithm, we usually do not use all three variables to form tree branches at each node. We use only m variables out of *M*. So we choose m such that m is less than or equal to *M*. The greater m is, the stronger the classifier is in each constructed tree. However, as mentioned earlier, more data leads to more bias. But, because we use multiple trees (with smaller *m*), even if each constructed tree is a weak classifier, their combined classification accuracy is strong. As we want to reduce a bias in a random forest, we may want to consider to choose a parameter m to be slightly less than *M*.

Thus we choose the maximum number of the variables considered at the node to be *m=min(M,math.ceil(2*math.sqrt(M)))=min(M,math.ceil(2*math.sqrt(3)))=3*.

We are given the following features:

```
[['None', 'Cold', 'No'], ['None', 'Warm', 'No'], ['Small', 'Cold', 'No'],
['Small', 'Warm', 'No'], ['Good', 'Cold', 'No'], ['Good', 'Warm', 'Yes']]
```

When constructing a random decision tree as a part of a random forest, we will choose only a subset out of them in a random manner with replacement.

Random forest construction

We construct a random forest that will consist of two random decision trees.

Construction of random decision tree number 0

We are given six features as the input data. Out of these, we choose randomly six features with replacement for the construction of this random decision tree:

```
[['None', 'Warm', 'No'], ['None', 'Warm', 'No'], ['Small', 'Cold', 'No'],
['Good', 'Cold', 'No'], ['Good', 'Cold', 'No'], ['Good', 'Cold', 'No']]
```

We start the construction with the root node to create the first node of the tree. We would like to add children to the node `[root]`.

We have the following variables available `['swimming_suit', 'water_temperature']`. As there are fewer of them than the parameter *m=3*, we consider all of them. Of these, the variable with the highest information gain is swimming suit.

Therefore, we will branch the node further on this variable. We also remove this variable from the list of the available variables for the children of the current node. Using the variable `swimming_suit`, we partition the data in the current node as follows:

- Partition for `swimming_suit=Small: [['Small', 'Cold', 'No']]`
- Partition for `swimming_suit=None: [['None', 'Warm', 'No'], ['None', 'Warm', 'No']]`
- Partition for `swimming_suit=Good: [['Good', 'Cold', 'No'], ['Good', 'Cold', 'No'], ['Good', 'Cold', 'No']]`

Using the preceding partitions, we create the branches and the child nodes.

We now add a child node `[swimming_suit=Small]` to the node `[root]`. This branch classifies one feature(s): `[['Small', 'Cold', 'No']]`.

We would like to add children to the node `[swimming_suit=Small]`.

We have the following variable available ['water_temperature']. As there are fewer of them than the parameter m=3, we consider all of them. Of these, the variable with the highest information gain is the variable water_temperature. Therefore, we will branch the node further on this variable. We also remove this variable from the list of the available variables for the children of the current node. For the chosen variable water_temperature, all the remaining features have the same value: Cold. So, we end the branch with a leaf node. We add the leaf node [swim=No].

We now add a child node [swimming_suit=None] to the node [root]. This branch classifies two feature(s): [['None', 'Warm', 'No'], ['None', 'Warm', 'No']].

We would like to add children to the node [swimming_suit=None].

We have the following variable available ['water_temperature']. As there are fewer of them than the parameter m=3, we consider all of them. Of these, the variable with the highest information gain is the variable water_temperature. Therefore, we will branch the node further on this variable. We also remove this variable from the list of the available variables for the children of the current node. For the chosen variable water_temperature, all the remaining features have the same value: Warm. So, we end the branch with a leaf node. We add the leaf node [swim=No].

We now add a child node [swimming_suit=Good] to the node [root]. This branch classifies three feature(s): [['Good', 'Cold', 'No'], ['Good', 'Cold', 'No'], ['Good', 'Cold', 'No']]

We would like to add children to the node [swimming_suit=Good].

We have the following variable available ['water_temperature']. As there are fewer of them than the parameter m=3, we consider all of them. Of these, the variable with the highest information gain is the variable water_temperature. Therefore, we will branch the node further on this variable. We also remove this variable from the list of the available variables for the children of the current node. For the chosen variable water_temperature, all the remaining features have the same value: Cold. So, we end the branch with a leaf node. We add the leaf node [swim=No].

Now, we have added all the children nodes for the node [root].

Construction of random decision tree number 1

We are given six features as the input data. Out of these, we choose randomly six features with replacement for the construction of this random decision tree:

```
[['Good', 'Warm', 'Yes'], ['None', 'Warm', 'No'], ['Good', 'Cold', 'No'],
['None', 'Cold', 'No'], ['None', 'Warm', 'No'], ['Small', 'Warm', 'No']]
```

The rest of the construction of random decision tree number 1 is similar to the construction of the previous random decision tree number 0. The only difference is that the tree is built with the different randomly generated subset (as seen above) of the initial data.

We start the construction with the root node to create the first node of the tree. We would like to add children to the node `[root]`.

We have the following variables available `['swimming_suit', 'water_temperature']`. As there are fewer of them than the parameter `m=3`, we consider all of them. Of these, the variable with the highest information gain is the variable `swimming_suit`.

Therefore, we will branch the node further on this variable. We also remove this variable from the list of the available variables for the children of the current node. Using the variable `swimming_suit`, we partition the data in the current node as follows:

- Partition for `swimming_suit=Small: [['Small', 'Warm', 'No']]`
- Partition for `swimming_suit=None: [['None', 'Warm', 'No'], ['None', 'Cold', 'No'], ['None', 'Warm', 'No']]`
- Partition for `swimming_suit=Good: [['Good', 'Warm', 'Yes'], ['Good', 'Cold', 'No']]`

Now, given the partitions, let us create the branches and the child nodes. We add a child node `[swimming_suit=Small]` to the node `[root]`. This branch classifies one feature(s): `[['Small', 'Warm', 'No']]`.

We would like to add children to the node `[swimming_suit=Small]`.

We have the following variable available `['water_temperature']`. As there are fewer of them than the parameter *m=3*, we consider all of them. Of these, the variable with the highest information gain is the variable `water_temperature`. Therefore, we will branch the node further on this variable. We also remove this variable from the list of the available variables for the children of the current node. For the chosen variable `water_temperature`, all the remaining features have the same value: Warm. So, we end the branch with a leaf node. We add the leaf node `[swim=No]`.

We add a child node `[swimming_suit=None]` to the node `[root]`. This branch classifies three feature(s): `[['None', 'Warm', 'No'], ['None', 'Cold', 'No'], ['None', 'Warm', 'No']]`.

We would like to add children to the node `[swimming_suit=None]`.

We have the following variable available `['water_temperature']`. As there are fewer of them than the parameter m=3, we consider all of them. Of these, the variable with the highest information gain is the variable `water_temperature`. Therefore, we will branch the node further on this variable. We also remove this variable from the list of the available variables for the children of the current node. Using the variable water temperature, we partition the data in the current node as follows:

- Partition for `water_temperature=Cold: [['None', 'Cold', 'No']]`
- Partition for `water_temperature=Warm: [['None', 'Warm', 'No'], ['None', 'Warm', 'No']]` Now, given the partitions, let us create the branches and the child nodes.

We add a child node `[water_temperature=Cold]` to the node `[swimming_suit=None]`. This branch classifies one feature(s): `[['None', 'Cold', 'No']]`.

We do not have any available variables on which we could split the node further; therefore, we add a leaf node to the current branch of the tree. We add the leaf node `[swim=No]`.

We add a child node `[water_temperature=Warm]` to the node `[swimming_suit=None]`. This branch classifies two feature(s): `[['None', 'Warm', 'No'], ['None', 'Warm', 'No']]`.

We do not have any available variables on which we could split the node further; therefore, we add a leaf node to the current branch of the tree. We add the leaf node `[swim=No]`.

Now, we have added all the children nodes for the node `[swimming_suit=None]`.

We add a child node `[swimming_suit=Good]` to the node `[root]`. This branch classifies two feature(s): `[['Good', 'Warm', 'Yes'], ['Good', 'Cold', 'No']]`

We would like to add children to the node `[swimming_suit=Good]`.

We have the following variable available ['water_temperature']. As there are fewer of them than the parameter *m=3*, we consider all of them. Out of these variables, the variable with the highest information gain is the variable water_temperature. Therefore, we will branch the node further on this variable. We also remove this variable from the list of the available variables for the children of the current node. Using the variable water temperature, we partition the data in the current node as follows:

- Partition for water_temperature=Cold: [['Good', 'Cold', 'No']]
- Partition for water_temperature=Warm: [['Good', 'Warm', 'Yes']]

Now, given the partitions, let us create the branches and the child nodes.

We add a child node [water_temperature=Cold] to the node [swimming_suit=Good]. This branch classifies one feature(s): [['Good', 'Cold', 'No']]

We do not have any available variables on which we could split the node further; therefore, we add a leaf node to the current branch of the tree. We add the leaf node [swim=No].

We add a child node [water_temperature=Warm] to the node [swimming_suit=Good]. This branch classifies one feature(s): [['Good', 'Warm', 'Yes']]

We do not have any available variables on which we could split the node further; therefore, we add a leaf node to the current branch of the tree. We add the leaf node [swim=Yes].

Now, we have added all the children nodes for the node [swimming_suit=Good].

Now, we have added all the children nodes for the node [root].

Therefore we have completed the construction of the random forest consisting of two random decision trees.

Random forest graph:

```
Tree 0:
    Root
    ├── [swimming_suit=Small]
    │  └── [swim=No]
    ├── [swimming_suit=None]
    │  └── [swim=No]
    └── [swimming_suit=Good]
       └── [swim=No]
Tree 1:
    Root
    ├── [swimming_suit=Small]
    │  └── [swim=No]
```

```
├── [swimming_suit=None]
│ ├── [water_temperature=Cold]
│ │ └── [swim=No]
│ └──[water_temperature=Warm]
│   └── [swim=No]
└── [swimming_suit=Good]
  ├──  [water_temperature=Cold]
  │ └── [swim=No]
  └──  [water_temperature=Warm]
    └── [swim=Yes]
The total number of trees in the random forest=2.
The maximum number of the variables considered at the node is m=3.
```

Classification with random forest

Because we use only a subset of the original data for the construction of the random decision tree, we may not have enough features to form a full tree that is able to classify every feature. In such cases, a tree will not return any class for a feature that should be classified. Therefore, we will only consider trees that classify a feature to some specific class.

The feature we would like to classify is: `['Good', 'Cold', '?']`. A random decision tree votes for the class to which it classifies a given feature using the same method to classify a feature as in the previous chapter on decision trees. Tree 0 votes for the class: No. Tree 1 votes for the class: No. The class with the maximum number of votes is 'No'. Therefore, the constructed random forest classifies the feature `['Good', 'Cold', '?']` into the class 'No'.

Implementation of random forest algorithm

We implement a random forest algorithm using a modified decision tree algorithm from the previous chapter. We also add an option to set a verbose mode within the program that can describe the whole process of how the algorithm works on a specific input- how a random forest is constructed with its random decision trees and how this constructed random forest is used to classify other features.

The implementation of a random forest uses the construction of a decision tree from the previous chapter. A reader is encouraged to consult the function `decision_tree.construct_general_tree` from the previous chapter:

```
# source_code/4/random_forest.py
import math
import random
```

```
import sys
sys.path.append('../common')
import common # noqa
import decision_tree # noqa
from common import printfv # noqa

#Random forest construction
def sample_with_replacement(population, size):
    sample = []
    for i in range(0, size):
        sample.append(population[random.randint(0, len(population) - 1)])
    return sample

def construct_random_forest(verbose, heading, complete_data,
                            enquired_column, m, tree_count):
    printfv(2, verbose, "*** Random Forest construction ***\n")
    printfv(2, verbose, "We construct a random forest that will " +
            "consist of %d random decision trees.\n", tree_count)
    random_forest = []
    for i in range(0, tree_count):
        printfv(2, verbose, "\nConstruction of a random " +
                "decision tree number %d:\n", i)
        random_forest.append(construct_random_decision_tree(
            verbose, heading, complete_data, enquired_column, m))
    printfv(2, verbose, "\nTherefore we have completed the " +
            "construction of the random forest consisting of %d " +
            "random decision trees.\n", tree_count)
    return random_forest

def construct_random_decision_tree(verbose, heading, complete_data,
                                   enquired_column, m):
    sample = sample_with_replacement(complete_data, len(complete_data))
    printfv(2, verbose, "We are given %d features as the input data. " +
            "Out of these, we choose randomly %d features with the " +
            "replacement that we will use for the construction of " +
            "this particular random decision tree:\n" +
            str(sample) + "\n", len(complete_data),
            len(complete_data))
# The function construct_general_tree from the module decision_tree
# is written in the implementation section in the previous chapter
# on decision trees.
    return decision_tree.construct_general_tree(verbose, heading,
                                                sample,
                                                enquired_column, m)

# M is the given number of the decision variables, i.e. properties
# of one feature.
def choose_m(verbose, M):
```

```
    m = int(min(M, math.ceil(2 * math.sqrt(M))))
    printfv(2, verbose, "We are given M=" + str(M) +
            " variables according to which a feature can be " +
            "classified. ")
    printfv(3, verbose, "In random forest algorithm we usually do " +
            "not use all " + str(M) + " variables to form tree " +
            "branches at each node. ")
    printfv(3, verbose, "We use only m variables out of M. ")
    printfv(3, verbose, "So we choose m such that m is less than or " +
            "equal to M. ")
    printfv(3, verbose, "The greater m is, a stronger classifier an " +
            "individual tree constructed is. However, it is more " +
            "susceptible to a bias as more of the data is considered. " +
            "Since we in the end use multiple trees, even if each may " +
            "be a weak classifier, their combined classification " +
            "accuracy is strong. Therefore as we want to reduce a " +
            "bias in a random forest, we may want to consider to " +
            "choose a parameter m to be slightly less than M.\n")
    printfv(2, verbose, "Thus we choose the maximum number of the " +
            "variables considered at the node to be " +
            "m=min(M,math.ceil(2*math.sqrt(M)))" +
            "=min(M,math.ceil(2*math.sqrt(%d)))=%d.\n", M, m)
    return m

#Classification
def display_classification(verbose, random_forest, heading,
                           enquired_column, incomplete_data):
    if len(incomplete_data) == 0:
        printfv(0, verbose, "No data to classify.\n")
    else:
        for incomplete_feature in incomplete_data:
            printfv(0, verbose, "\nFeature: " +
                    str(incomplete_feature) + "\n")
            display_classification_for_feature(
                verbose, random_forest, heading,
                enquired_column, incomplete_feature)

def display_classification_for_feature(verbose, random_forest, heading,
                                       enquired_column, feature):
    classification = {}
    for i in range(0, len(random_forest)):
        group = decision_tree.classify_by_tree(
            random_forest[i], heading, enquired_column, feature)
        common.dic_inc(classification, group)
        printfv(0, verbose, "Tree " + str(i) +
                " votes for the class: " + str(group) + "\n")
    printfv(0, verbose, "The class with the maximum number of votes " +
            "is '" + str(common.dic_key_max_count(classification)) +
```

```
        "'. Thus the constructed random forest classifies the " +
        "feature " + str(feature) + " into the class '" +
        str(common.dic_key_max_count(classification)) + "'.\n")

# Program start
csv_file_name = sys.argv[1]
tree_count = int(sys.argv[2])
verbose = int(sys.argv[3])

(heading, complete_data, incomplete_data,
 enquired_column) = common.csv_file_to_ordered_data(csv_file_name)
m = choose_m(verbose, len(heading))
random_forest = construct_random_forest(
    verbose, heading, complete_data, enquired_column, m, tree_count)
display_forest(verbose, random_forest)
display_classification(verbose, random_forest, heading,
                       enquired_column, incomplete_data)
```

Input:

As an input file to the implemented algorithm we provide the data from example Swim preference.

```
# source_code/4/swim.csv
swimming_suit,water_temperature,swim
None,Cold,No
None,Warm,No
Small,Cold,No
Small,Warm,No
Good,Cold,No
Good,Warm,Yes
Good,Cold,?
```

Output:

We type the following command in the command line to get the output:

```
$ python random_forest.py swim.csv 2 3 > swim.out
```

`2` means that we would like to construct 2 decision trees and `3` is the level of the verbosity of the program which includes detailed explanations of the construction of the random forest, the classification of the feature and the graph of the random forest. The last part `> swim.out` means that the output is written to the file `swim.out`. This file can be found in the chapter directory `source_code/4`. This output of the program was used above to write the analysis of Swim preference problem.

Playing chess example

We will use the example from the `Chapter 2`, *Naive Bayes* and `Chapter 3`, *Decision Tree,* again.

Temperature	Wind	Sunshine	Play
Cold	Strong	Cloudy	No
Warm	Strong	Cloudy	No
Warm	None	Sunny	Yes
Hot	None	Sunny	No
Hot	Breeze	Cloudy	Yes
Warm	Breeze	Sunny	Yes
Cold	Breeze	Cloudy	No
Cold	None	Sunny	Yes
Hot	Strong	Cloudy	Yes
Warm	None	Cloudy	Yes
Warm	Strong	Sunny	?

However, we would like to use a random forest consisting of four random decision trees to find the result of the classification.

Analysis:

We are given *M=4* variables from which a feature can be classified. Thus, we choose the maximum number of the variables considered at the node to be *m=min(M,math.ceil(2*math.sqrt(M)))=min(M,math.ceil(2*math.sqrt(4)))=4.*

We are given the following features:

```
[['Cold', 'Strong', 'Cloudy', 'No'], ['Warm', 'Strong', 'Cloudy', 'No'],
['Warm', 'None', 'Sunny',
'Yes'], ['Hot', 'None', 'Sunny', 'No'], ['Hot', 'Breeze', 'Cloudy', 'Yes'],
['Warm', 'Breeze',
'Sunny', 'Yes'], ['Cold', 'Breeze', 'Cloudy', 'No'], ['Cold', 'None',
'Sunny', 'Yes'], ['Hot', 'Strong', 'Cloudy', 'Yes'], ['Warm', 'None',
'Cloudy', 'Yes']]
```

When constructing a random decision tree as a part of a random forest, we will choose only a subset of them in a random way with replacement.

Random forest construction

We construct a random forest that will consist of four random decision trees.

Construction of a random decision tree number 0:

We are given 10 features as the input data. Out of these, we choose randomly 10 features with replacement that we will use for the construction of this random decision tree:

```
[['Warm', 'Strong', 'Cloudy', 'No'], ['Cold', 'Breeze', 'Cloudy', 'No'],
['Cold', 'None', 'Sunny', 'Yes'], ['Cold', 'Breeze', 'Cloudy', 'No'],
['Hot', 'Breeze', 'Cloudy', 'Yes'], ['Warm', 'Strong', 'Cloudy', 'No'],
['Hot', 'Breeze', 'Cloudy', 'Yes'], ['Hot', 'Breeze', 'Cloudy', 'Yes'],
['Cold', 'Breeze', 'Cloudy', 'No'], ['Warm', 'Breeze', 'Sunny', 'Yes']]
```

We start the construction with the root node to create the first node of the tree. We would like to add children to the node [root].

We have the following variables available `['Temperature', 'Wind', 'Sunshine']`. As there are fewer of them than the parameter *m=4*, we consider all of them. Of these, the variable with the highest information gain is the variable Temperature. Therefore, we will branch the node further on this variable. We also remove this variable from the list of the available variables for the children of the current node. Using the variable Temperature, we partition the data in the current node as follows:

- Partition for `Temperature=Cold: [['Cold', 'Breeze', 'Cloudy', 'No'], ['Cold', 'None', 'Sunny', 'Yes'], ['Cold', 'Breeze', 'Cloudy', 'No'], ['Cold', 'Breeze', 'Cloudy', 'No']]`
- Partition for `Temperature=Warm: [['Warm', 'Strong', 'Cloudy', 'No'], ['Warm', 'Strong', 'Cloudy', 'No'], ['Warm', 'Breeze', 'Sunny', 'Yes']]`
- Partition for `Temperature=Hot: [['Hot', 'Breeze', 'Cloudy', 'Yes'], ['Hot', 'Breeze', 'Cloudy', 'Yes'], ['Hot', 'Breeze', 'Cloudy', 'Yes']]`

Now, given the partitions, let us create the branches and the child nodes.

We add a child node `[Temperature=Cold]` to the node `[root]`. This branch classifies four feature(s): `[['Cold', 'Breeze', 'Cloudy', 'No'], ['Cold', 'None', 'Sunny', 'Yes'], ['Cold', 'Breeze', 'Cloudy', 'No'], ['Cold', 'Breeze', 'Cloudy', 'No']]`.

We would like to add children to the node `[Temperature=Cold]`.

We have the following variables available `['Wind', 'Sunshine']`. As there are fewer of them than the parameter *m=4*, we consider all of them. Of these, the variable with the highest information gain is the variable Wind. Therefore, we will branch the node further on this variable. We also remove this variable from the list of the available variables for the children of the current node. Using the variable water Wind, we partition the data in the current node as follows:

- Partition for `Wind=None: [['Cold', 'None', 'Sunny', 'Yes']]`
- Partition for `Wind=Breeze: [['Cold', 'Breeze', 'Cloudy', 'No'], ['Cold', 'Breeze', 'Cloudy', 'No'], ['Cold', 'Breeze', 'Cloudy', 'No']]`

Now, given the partitions, let us create the branches and the child nodes.

We add a child node `[Wind=None]` to the node `[Temperature=Cold]`. This branch classifies one feature(s): `[['Cold', 'None', 'Sunny', 'Yes']]`

We would like to add children to the node `[Wind=None]`.

We have the following variable `available['Sunshine']`. As there are fewer of them than the parameter `m=4`, we consider all of them. Of these, the variable with the highest information gain is the variable Sunshine. Therefore, we will branch the node further on this variable. We also remove this variable from the list of the available variables for the children of the current node. For the chosen variable Sunshine, all the remaining features have the same value: Sunny. So, we end the branch with a leaf node. We add the leaf node `[Play=Yes]`.

We add a child node `[Wind=Breeze]` to the node `[Temperature=Cold]`. This branch classifies three feature(s): `[['Cold', 'Breeze', 'Cloudy', 'No'], ['Cold', 'Breeze', 'Cloudy', 'No'], ['Cold', 'Breeze', 'Cloudy', 'No']]`

We would like to add children to the node `[Wind=Breeze]`.

We have the following variable available `['Sunshine']`. As there are fewer of them than the parameter *m=4*, we consider all of them. Of these, the variable with the highest information gain is the variable Sunshine. Therefore, we will branch the node further on this variable. We also remove this variable from the list of the available variables for the children of the current node. For the chosen variable Sunshine, all the remaining features have the same `value: Cloudy`. So, we end the branch with a leaf node. We add the leaf node `[Play=No]`.

Now, we have added all the children nodes for the node `[Temperature=Cold]`.

We add a child node `[Temperature=Warm]` to the node `[root]`. This branch classifies three feature(s): `[['Warm', 'Strong', 'Cloudy', 'No'], ['Warm', 'Strong', 'Cloudy', 'No'], ['Warm', 'Breeze', 'Sunny', 'Yes']]`

We would like to add children to the node [Temperature=Warm].

The available variables that we have still left are `['Wind', 'Sunshine']`. As there are fewer of them than the parameter *m=4*, we consider all of them. Out of these variables, the variable with the highest information gain is the variable Wind. Thus we will branch the node further on this variable. We also remove this variable from the list of the available variables for the children of the current node. Using the variable Wind, we partition the data in the current node, where each partition of the data will be for one of the new branches from the current node [Temperature=Warm]. We have the following partitions:

- Partition for `Wind=Breeze: [['Warm', 'Breeze', 'Sunny', 'Yes']]`
- Partition for `Wind=Strong: [['Warm', 'Strong', 'Cloudy', 'No'], ['Warm', 'Strong', 'Cloudy', 'No']]`

Now, given the partitions, let us form the branches and the child nodes.

We add a child node `[Wind=Breeze]` to the node `[Temperature=Warm]`. This branch classifies one feature(s): `[['Warm', 'Breeze', 'Sunny', 'Yes']]`

We would like to add children to the node `[Wind=Breeze]`.

We have the following variable available ['Sunshine']. As there are fewer of them than the parameter *m=4*, we consider all of them. Of these, the variable with the highest information gain is the variable Sunshine. Therefore, we will branch the node further on this variable. We also remove this variable from the list of the available variables for the children of the current node. For the chosen variable Sunshine, all the remaining features have the same value: Sunny. So, we end the branch with a leaf node. We add the leaf node [Play=Yes].

We add a child node [Wind=Strong] to the node [Temperature=Warm]. This branch classifies two feature(s): [['Warm', 'Strong', 'Cloudy', 'No'], ['Warm', 'Strong', 'Cloudy', 'No']]

We would like to add children to the node [Wind=Strong].

We have the following variable available ['Sunshine']. As there are fewer of them than the parameter *m=4*, we consider all of them. Of these, the variable with the highest information gain is the variable: Sunshine. Therefore, we will branch the node further on this variable. We also remove this variable from the list of the available variables for the children of the current node. For the chosen variable Sunshine, all the remaining features have the same value: Cloudy. So, we end the branch with a leaf node. We add the leaf node [Play=No].

Now, we have added all the children nodes for the node [Temperature=Warm].

We add a child node [Temperature=Hot] to the node [root]. This branch classifies three feature(s): [['Hot', 'Breeze', 'Cloudy', 'Yes'], ['Hot', 'Breeze', 'Cloudy', 'Yes'], ['Hot', 'Breeze', 'Cloudy', 'Yes']]

We would like to add children to the node [Temperature=Hot].

We have the following variables available ['Wind', 'Sunshine']. As there are fewer of them than the parameter *m=4*, we consider all of them. Of these, the variable with the highest information gain is the variable Wind. Therefore, we will branch the node further on this variable. We also remove this variable from the list of the available variables for the children of the current node. For the chosen variable Wind, all the remaining features have the same value: Breeze. So, we end the branch with a leaf node. We add the leaf node [Play=Yes].

Now, we have added all the children nodes for the node [root].

Construction of a random decision tree number 1, 2, 3

We construct the next three trees in a similar fashion. We should note that since the construction is random, a reader who performs another correct construction may arrive at a different construction. However, if there are sufficiently many random decision trees in a random forest, then the result of the classification should be very similar across all the random constructions.

The full construction can be found in the program output in the file `source_code/4/chess.out`.

Random forest graph:

```
Tree 0:
    Root
    ├── [Temperature=Cold]
    | ├── [Wind=None]
    | | └── [Play=Yes]
    | └──[Wind=Breeze]
    |   └── [Play=No]
    ├── [Temperature=Warm]
    | ├──[Wind=Breeze]
    | | └── [Play=Yes]
    | └──[Wind=Strong]
    |   └── [Play=No]
    └── [Temperature=Hot]
        └── [Play=Yes]
Tree 1: Root ├── [Wind=Breeze] | └── [Play=No] ├── [Wind=None] | ├──
[Temperature=Cold] | | └── [Play=Yes] | ├── [Temperature=Warm] | | ├──
[Sunshine=Sunny] | | | └──[Play=Yes] | | └──[Sunshine=Cloudy] | | └──
[Play=Yes] | └── [Temperature=Hot] | └── [Play=No] └── [Wind=Strong] ├──
[Temperature=Cold] | └── [Play=No] └── [Temperature=Warm] └── [Play=No]
Tree 2:
    Root
    ├── [Wind=Strong]
    | └── [Play=No]
    ├── [Wind=None]
    | ├── [Temperature=Cold]
    | | └── [Play=Yes]
    | └── [Temperature=Warm]
    |   └── [Play=Yes]
    └── [Wind=Breeze]
      ├── [Temperature=Hot]
      | └── [Play=Yes]
      └── [Temperature=Warm]
        └── [Play=Yes]
```

```
Tree 3:
    Root
    ├── [Temperature=Cold]
    │ └── [Play=No]
    ├── [Temperature=Warm]
    │ ├──[Wind=Strong]
    │ │ └──[Play=No]
    │ ├── [Wind=None]
    │ │ └── [Play=Yes]
    │ └──[Wind=Breeze]
    │   └── [Play=Yes]
    └── [Temperature=Hot]
      ├── [Wind=Strong]
      │ └── [Play=Yes]
      └── [Wind=Breeze]
        └── [Play=Yes]
The total number of trees in the random forest=4.
The maximum number of the variables considered at the node is m=4.
```

Classification:

Given the constructed random forest we classify feature `['Warm', 'Strong', 'Sunny', '?']`:

- Tree 0 votes for the class: No
- Tree 1 votes for the class: No
- Tree 2 votes for the class: No
- Tree 3 votes for the class: No

The class with the maximum number of votes is 'No'. Thus the constructed random forest classifies the feature `['Warm', 'Strong', 'Sunny', '?']` into the class 'No'.

Input:

To perform the preceding analysis, we use a program implemented earlier in this chapter. First we put the data from the table into the following CSV file:

```
# source_code/4/chess.csv
Temperature,Wind,Sunshine,Play
Cold,Strong,Cloudy,No
Warm,Strong,Cloudy,No
Warm,None,Sunny,Yes
Hot,None,Sunny,No
Hot,Breeze,Cloudy,Yes
Warm,Breeze,Sunny,Yes
Cold,Breeze,Cloudy,No
```

```
Cold,None,Sunny,Yes
Hot,Strong,Cloudy,Yes
Warm,None,Cloudy,Yes
Warm,Strong,Sunny,?
```

Output:

We produce the output by executing on the command line:

```
$ python random_forest.py chess.csv 4 2 > chess.out
```

The number `4` here means that we want to construct four decision trees and `2` is the level of the verbosity of the program which includes the explanations of a tree is constructed. The last part `> chess.out` means that the output is written to the file `chess.out`. This file can be found in the chapter directory `source_code/4`. We do not put all the output here, as it is very large and repetitive. Instead some of it was included in the preceding analysis and construction of a random forest.

Going shopping - overcoming data inconsistency with randomness and measuring the level of confidence

We take the problem from the previous chapter. We have the following data about the shopping preferences of our friend, Jane:

Temperature	Rain	Shopping
Cold	None	Yes
Warm	None	No
Cold	Strong	Yes
Cold	None	No
Warm	Strong	No
Warm	None	Yes
Cold	None	?

In the previous chapter, decision trees were not able to classify the feature `(Cold, None)`. So, this time, we would like to find, using the random forest algorithm, whether Jane would go shopping if the outside temperature was cold and there was no rain.

Analysis:

To perform the analysis with the random forest algorithm we use the implemented program.

Input:

We put the data from the table into the CSV file:

```
# source_code/4/shopping.csv
Temperature,Rain,Shopping
Cold,None,Yes
Warm,None,No
Cold,Strong,Yes
Cold,None,No
Warm,Strong,No
Warm,None,Yes
Cold,None,?
```

Output:

We want to use a slightly larger number of the trees that we used in the previous examples and explanations to get more accurate results. We want to construct a random forest with 20 trees with the output of the low verbosity - level 0. Thus, we execute in a terminal:

```
$ python random_forest.py shopping.csv 20 0
***Classification***
Feature: ['Cold', 'None', '?']
Tree 0 votes for the class: Yes
Tree 1 votes for the class: No
Tree 2 votes for the class: No
Tree 3 votes for the class: No
Tree 4 votes for the class: No
Tree 5 votes for the class: Yes
Tree 6 votes for the class: Yes
Tree 7 votes for the class: Yes
Tree 8 votes for the class: No
Tree 9 votes for the class: Yes
Tree 10 votes for the class: Yes
Tree 11 votes for the class: Yes
Tree 12 votes for the class: Yes
Tree 13 votes for the class: Yes
Tree 14 votes for the class: Yes
```

```
Tree 15 votes for the class: Yes
Tree 16 votes for the class: Yes
Tree 17 votes for the class: No
Tree 18 votes for the class: No
Tree 19 votes for the class: No
The class with the maximum number of votes is 'Yes'. Thus the constructed
random forest classifies the feature ['Cold', 'None', '?'] into the class
'Yes'.
```

However, we should note that only 12 out of the 20 trees voted for the answer *Yes*. Thus just as an ordinary decision tree could not decide the case, so here, although having a definite answer, it may not be so certain. But unlike in decision trees where an answer was not produced because of data inconsistency, here we have an answer.

Furthermore, by measuring the strength of the voting power for each individual class, we can measure the level of the confidence that the answer is correct. In this case the feature `['Cold', 'None', '?']` belongs to the class *Yes* with the confidence of 12/20 or 60%. To determine the level of certainty of the classification more precisely, even a larger ensemble of random decision trees would be required.

Summary

A random forest is a set of decision trees where each tree is constructed from a sample chosen randomly from the initial data. This process is called bootstrap aggregating. Its purpose is to reduce variance and bias in the classification made by a random forest. The bias is further reduced during a construction of a decision tree by considering only a random subset of the variables for each branch of the tree.

Once a random forest is constructed, the result of the classification of a random forest is the majority vote from among all the trees in a random forest. The level of the majority also determines the amount of the confidence that the answer is correct.

Since a random forest consists of decision trees, it is good to use it for every problem where a decision tree is a good choice. Since a random forest reduces bias and variance that exist in a decision tree classifier, it outperforms a decision tree algorithm.

Problems

Let us take another example of playing chess from `Chapter 2`, *Naive Bayes*. How would you classify a data sample (warm,strong,spring,?) according to the random forest algorithm?

Temperature	Wind	Season	Play
Cold	Strong	Winter	No
Warm	Strong	Autumn	No
Warm	None	Summer	Yes
Hot	None	Spring	No
Hot	Breeze	Autumn	Yes
Warm	Breeze	Spring	Yes
Cold	Breeze	Winter	No
Cold	None	Spring	Yes
Hot	Strong	Summer	Yes
Warm	None	Autumn	Yes
Warm	Strong	Spring	?

Would it be a good idea to use only one tree and a random forest? Justify your answer.

Can cross-validation improve the results of the classification by the random forest? Justify your answer.

Analysis:

1. We run the program to construct the random forest and classify the feature (Warm, Strong, Spring).

Input:

source_code/4/chess_with_seasons.csv

```
Temperature,Wind,Season,Play
Cold,Strong,Winter,No
Warm,Strong,Autumn,No
Warm,None,Summer,Yes
Hot,None,Spring,No
Hot,Breeze,Autumn,Yes
Warm,Breeze,Spring,Yes
```

```
Cold,Breeze,Winter,No
Cold,None,Spring,Yes
Hot,Strong,Summer,Yes
Warm,None,Autumn,Yes
Warm,Strong,Spring,?
```

Output:

We construct four trees in a random forest:

```
$ python chess_with_seasons.csv 4 2 > chess_with_seasons.out
```

The whole construction and analysis is stored in the file `source_code/4/chess_with_seasons.out`. Your construction may differ because of the randomness involved. From the output we extract the random forest graph consisting of random decision trees given the random numbers generated during our run.

Executing the command above again will most likely result in a different output and different random forest graph. Yet the results of the classification should be similar with a high probability because of the multiplicity of the random decision trees and their voting power combined. The classification by one random decision tree may be subject to a great variance. However, the majority vote combines the classification from all the trees, thus reducing the variance. To verify your understanding, you can compare your results of the classification with the classification by the random forest graph below.

Random forest graph and classification:

Let's have a look at the output of the random forest graph and the classification of the feature:

```
Tree 0:
    Root
    ├── [Wind=None]
    │  ├── [Temperature=Cold]
    │  │  └── [Play=Yes]
    │  └── [Temperature=Warm]
    │     ├── [Season=Autumn]
    │     │  └── [Play=Yes]
    │     └── [Season=Summer]
    │        └── [Play=Yes]
    └── [Wind=Strong]
       ├── [Temperature=Cold]
       │  └── [Play=No]
       └── [Temperature=Warm]
```

```
        └── [Play=No]
Tree 1:
    Root
    ├── [Season=Autumn]
    │  ├──[Wind=Strong]
    │  │  └──[Play=No]
    │  ├── [Wind=None]
    │  │  └── [Play=Yes]
    │  └──[Wind=Breeze]
    │     └── [Play=Yes]
    ├── [Season=Summer]
    │     └── [Play=Yes]
    ├── [Season=Winter]
    │     └── [Play=No]
    └── [Season=Spring]
       ├── [Temperature=Cold]
       │  └── [Play=Yes]
       └── [Temperature=Warm]
          └── [Play=Yes]
Tree 2:
    Root
    ├── [Season=Autumn]
    │  ├── [Temperature=Hot]
    │  │  └── [Play=Yes]
    │  └── [Temperature=Warm]
    │     └── [Play=No]
    ├── [Season=Spring]
    │  ├── [Temperature=Cold]
    │  │  └── [Play=Yes]
    │  └── [Temperature=Warm]
    │     └── [Play=Yes]
    ├── [Season=Winter]
    │  └── [Play=No]
    └── [Season=Summer]
       ├── [Temperature=Hot]
       │  └── [Play=Yes]
       └── [Temperature=Warm]
          └── [Play=Yes]
Tree 3:
    Root
    ├── [Season=Autumn]
    │  ├──[Wind=Breeze]
    │  │  └── [Play=Yes]
    │  ├── [Wind=None]
    │  │  └── [Play=Yes]
    │  └──[Wind=Strong]
    │     └── [Play=No]
    ├── [Season=Spring]
```

```
    │ ├── [Temperature=Cold]
    │ │ └── [Play=Yes]
    │ └── [Temperature=Warm]
    │   └── [Play=Yes]
    ├── [Season=Winter]
    │ └── [Play=No]
    └── [Season=Summer]
      └── [Play=Yes]
The total number of trees in the random forest=4.
The maximum number of the variables considered at the node is m=4.
Classification
Feature: ['Warm', 'Strong', 'Spring', '?']
Tree 0 votes for the class: No
Tree 1 votes for the class: Yes
Tree 2 votes for the class: Yes
Tree 3 votes for the class: Yes
The class with the maximum number of votes is 'Yes'. Thus the constructed
random forest classifies the feature ['Warm', 'Strong', 'Spring', '?'] into
the class 'Yes'.
```

When we construct a tree in a random forest, we use only a random subset of the data with replacement. This is to eliminate the bias of the classifier towards certain features. However, if we use only one tree, that tree may happen to contain features with bias and might miss some important feature to provide an accurate classification. So, a random forest classifier with one decision tree would likely lead to a very poor classification. Therefore, we should construct more decision trees in a random forest to benefit from the reduction of bias and variance in the classification.

During cross-validation, we divide the data into the training and the testing data. Training data is used to train the classifier and the test data is to evaluate which parameters or methods would be the best fit to improve the classification. Another advantage of cross-validation is the reduction of bias because we only use partial data, thereby decreasing the chance of overfitting to the specific dataset.

However, in a decision forest, we address problems that cross-validation addresses in an alternative way. Each random decision tree is constructed only on the subset of the data - reducing the chance of overfitting. In the end, the classification is the combination of results from each of these trees. The best decision in the end is not made by tuning the parameters on a test dataset, but by taking the majority vote of all the trees with reduced bias.

Hence, cross-validation for a decision forest algorithm would not be of a much use as it is already intrinsic within the algorithm.

5
Clustering into K Clusters

Clustering is a technique to divide the data into clusters so that features in the same cluster are in a certain sense similar.

In this chapter you will learn:

- The k-means clustering algorithm on example about household incomes
- An example about gender classification to classify features by clustering them first with the features with the known classes
- To implement k-means clustering algorithm in Python in section *Implementation of k-means clustering algorithm*
- An example about house ownership and how to choose an appropriate number of clusters for your analysis
- Using the example about house ownership how to scale given data appropriately to improve the accuracy of the classification by a clustering algorithm
- An example about document clustering to understand how a different number of clusters alters the meaning of the dividing boundary between the clusters

Household incomes - clustering into k clusters

For example let us take households with the yearly earnings in USD dollars 40k, 55k, 70k, 100k, 115k, 130k, 135k. Then if we require to cluster the households into the two clusters taking their earnings as a measure of similarity, then the first cluster would have the households earning 40k, 55k, 70k; the second cluster would have the households earning 100k, 115k, 130k, 135k.

This is because 40k and 135k are furthest away from each other, and we require to have two clusters, so they have to be in the different clusters. 55K is closer to 40k than to 135k, so 40k and 55k will be in the same cluster. Similarly, 130k and 135k will be in the same cluster. 70K is closer to 40k and 55k than to 130k and 135k, so 70k should be in the cluster with 40k and 55k. 115K is closer to 130k and 135k than to the first cluster with 40k, 55k and 70k, so it will be in the second cluster. Finally, 100k is closer to the second cluster with 115k, 130k and 135k, so it will be there. Therefore the first cluster will contain 40k, 55k and 70k households. The second cluster will contain 100k, 115k, 130k and 135k households.

Clustering groups features with similar properties and assigning a cluster to a feature is a form of classification. It is up to a data scientist to interpret the result of the clustering and what classification it induces. Here the cluster with the households with the annual incomes 40k, 55k, 70k USD represents a class of households with a low income. The second cluster with the households with the annual incomes 100k, 115k, 130k and 135k represents a class of households with a high income.

We clustered the households into the two clusters in an informal way based on the intuition and the common sense. There are clustering algorithms that cluster the data according to the precise rules. The algorithms include fuzzy c-means clustering algorithm, hierarchical clustering algorithm, Gaussian(EM) clustering algorithm, Quality Threshold clustering algorithm and k-means clustering algorithm which is the focus of this chapter.

K-means clustering algorithm

The k-means clustering algorithm classifies given points into k groups in such a way that a distance between the members of the same group is minimized.

The k-means clustering algorithm determines the initial k-centroids (points to be in a cluster center) – one for each cluster. Then each feature is classified into the cluster whose centroid is closest to that feature. After classifying all the features, we have formed initial k clusters.

For each cluster we recompute the centroid to be the average of the points in that cluster. After we have moved the centroids, we recompute the classes again. Features may change the classes. Then we will have to recompute the centroids again. If the centroids do not move anymore, then the k-means clustering algorithm terminates.

Picking the initial k-centroids

We could pick up the initial k-centroids to be any of the k features in the data to be classified. But ideally, we would like to pick up the points that belong to the different clusters already in the beginning. Therefore we may want to aim to maximize their mutual distance in a certain way. Simplifying the process we could pick the first centroid to be any point from the features. The second could be the one which is furthest from the first. The third could be the one that is furthest from both first and second, and so on.

Computing a centroid of a given cluster

A centroid of a cluster is just an average of the points in a cluster. If a cluster contains 1 dimensional points with the coordinates $x_1, x_2, \ldots, x_n$, then the centroid of that cluster would be $(1/n)*(x_1+x_2+\ldots+x_n)$. If a cluster contains 2 dimensional points with the coordinates $(x_1,y_1),(x_2,y_2),\ldots,(x_n,y_n)$, then the x coordinate of the centroid of the cluster would have value $(1/n)*(x_1+x_2+\ldots+x_n)$, the y coordinate would have the value $(1/n)*(y_1+y_2+\ldots+y_n)$.

This computation generalizes easily to higher dimensions. If the value of the higher dimensional features for the x-coordinate are $x_1, x_2, \ldots, x_n$, then the value at the x-coordinate for the centroid is $(1/n)*(x_1+x_2+\ldots+x_n)$.

k-means clustering algorithm on household income example

We will apply k-clustering algorithm on the household income example. In the beginning we have households with the incomes 40k, 55k, 70k, 100k, 115k, 130k and 135k in USD dollars.

The first centroid to be picked up can be any feature, example 70k. The second centroid should be the feature that is furthest from the first one, that is 135k since 135k-70k is 65k which is the greatest difference between any other feature and 70k. Thus 70k is the centroid of the first cluster, 135k is the centroid of the second cluster.

Now 40k, 55k, 70k, 100k are closer to 70k by taking the difference than to 135k, so they will be in the first cluster. The features 115k, 130k and 135k are closer to 135k than to 70k, so they will be in the second cluster.

After we have classified the features according to the initial centroids, we recompute the centroids. The centroid of the first cluster is $(1/4)*(40k+55k+70k+100k)=(1/4)*265k=66.25k$.

The centroid of the second cluster is *(1/3)*(115k+130k+135k)=(1/3)*380k~126.66k*.

Using the new centroids we reclassify the features as follows:

- The first cluster with the centroid 66.25k will contain the features 40k, 55k, 70k.
- The second cluster with the centroid 126.66k will contain the features 100k, 115k, 130k, 135k.

We notice that the feature 100k moved from the first cluster into the second since now it is closer to the centroid of the second cluster (distance |100k-126.66k|=26.66k) than to the centroid of the first cluster (distance |100k-66.25k|=33.75k). Since the features in the clusters changed, we have to recompute the centroids again.

The centroid of the first cluster is *(1/3)*(40k+55k+70k)=(1/3)/165k=55k*. The centroid of the second cluster is *(1/4)*(100k+115k+130k+135k)=(1/4)*480k=120k*.

Using these centroids we reclassify the items into the clusters. The first centroid 55k will contain the features 40k, 55k, 70k. The second centroid 120k will contain the features 100k, 115k, 130k, 135k. Thus upon the update of the centroids, the clusters did not change. So their centroids will remain the same.

Therefore the algorithm terminates with the two clusters: the first cluster having the features 40k, 55k, 70k; the second cluster having the features 100k, 115k, 130k, 135k.

Gender classification - clustering to classify

We take the data from the gender classification in the problem `Chapter 2`, *Naive Bayes*, Analysis point 6:

Height in cm	Weight in kg	Hair length	Gender
180	75	Short	Male
174	71	Short	Male
184	83	Short	Male
168	63	Short	Male
178	70	Long	Male
170	59	Long	Female
164	53	Short	Female

155	46	Long	Female
162	52	Long	Female
166	55	Long	Female
172	60	Long	?

To simplify the matters we will remove the column Hair length. We also remove the column Gender since we would like to cluster the people in the table based on their height and weight. We would like to find out whether the 11th person in the table is more likely to be a man or a woman using clustering:

Height in cm	**Weight in kg**
180	75
174	71
184	83
168	63
178	70
170	59
164	53
155	46
162	52
166	55
172	60

Analysis:

We may apply scaling to the initial data, but to simplify the matters, we will use the unscaled data in the algorithm. We will cluster the data we have into the two clusters since there are two possibilities for genders – a male or a female. Then we will aim to classify a person with the height 172cm and weight 60kg to be more likely a man if and only if there are more men in that cluster. The clustering algorithm is a very efficient technique. Thus classifying this way is very fast, especially if there is a large number of the features to classify.

So let us apply k-means clustering algorithm to the data we have. First we pick up the initial centroids. Let the first centroid be for example a person with the height 180cm and the weight 75kg denoted in a vector as *(180,75)*. Then the point that is furthest away from *(180,75)* is *(155,46)*. So that will be the second centroid.

The points that are closer to the first centroid *(180,75)* by taking Euclidean distance are *(180,75)*, *(174,71)*, *(184,83)*, *(168,63)*, *(178,70)*, *(170,59)*, *(172,60)*. So these points will be in the first cluster. The points that are closer to the second centroid *(155,46)* are *(155,46)*, *(164,53)*, *(162,52)*, *(166,55)*. So these points will be in the second cluster. We display the current situation of these two clusters in Image 5.1. below.

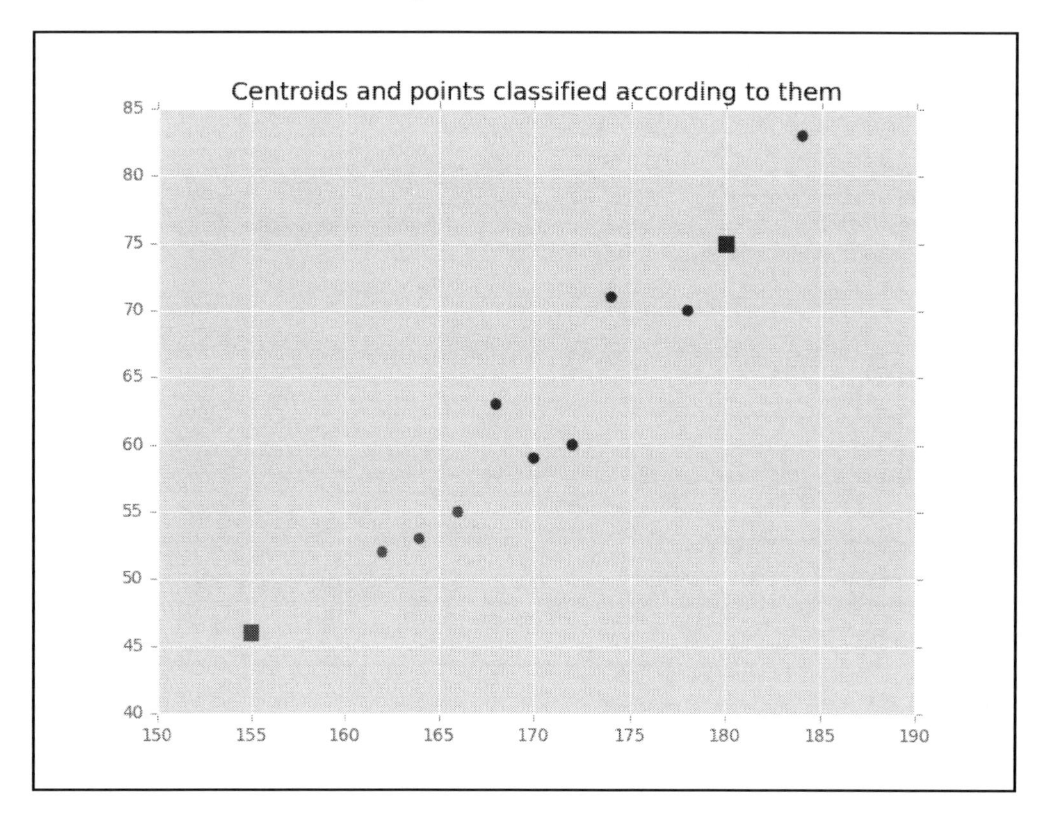

Image 5.1: Clustering of people by their height and weight

Let us recompute the centroids of the clusters. The blue cluster with the features *(180,75)*, *(174,71)*, *(184,83)*, *(168,63)*, *(178,70)*, *(170,59)*, *(172,60)* will have the centroid *((180+174+184+168+178+170+172)/7,(75+71+83+63+70+59+60)/7)~(175.14,68.71)*.

The red cluster with the features *(155,46), (164,53), (162,52), (166,55)* will have the centroid *((155+164+162+166)/4,(46+53+52+55)/4)=(161.75, 51.5).*

Reclassifying the points using the new centroid, the classes of the points do not change. The blue cluster will have the points (180,75), (174,71), (184,83), (168,63), (178,70), (170,59), (172,60). The red cluster will have the points (155,46), (164,53), (162,52), (166,55). Therefore the clustering algorithm terminates with clusters as displayed in the following image 5.2:

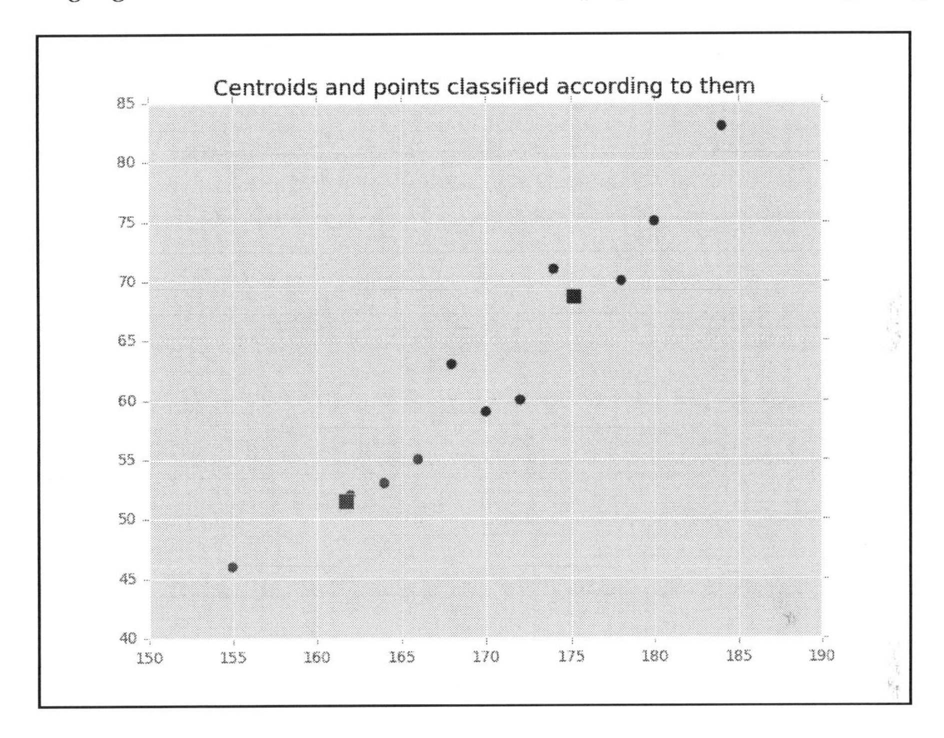

Image 5.2: Clustering of people by their height and weight

Now we would like to classify the instance *(172,60)* as to whether it is a male or a female. The instance (172,60) is in the blue cluster. So it is similar to the features in the blue cluster. Are the remaining features in the blue cluster more likely males or females? 5 out of 6 features are males, only 1 is a female. Since the majority of the features are males in the blue cluster and the person (172,60) is in the blue cluster as well, we classify the person with the height 172cm and the weight 60kg as a male.

Implementation of the k-means clustering algorithm

We implement the k-means clustering algorithm. It takes as an input a CSV file with one data item per line. A data item is converted to a point. The algorithms classifies these points into the specified number of clusters. In the end the clusters are visualized on the graph using the library `matplotlib`:

```
# source_code/5/k-means_clustering.py
import math
import imp
import sys
import matplotlib.pyplot as plt
import matplotlib
import sys
sys.path.append('../common')
import common # noqa
matplotlib.style.use('ggplot')

# Returns k initial centroids for the given points.
def choose_init_centroids(points, k):
    centroids = []
    centroids.append(points[0])
    while len(centroids) < k:
        # Find the centroid that with the greatest possible distance
        # to the closest already chosen centroid.
        candidate = points[0]
        candidate_dist = min_dist(points[0], centroids)
        for point in points:
            dist = min_dist(point, centroids)
            if dist > candidate_dist:
                candidate = point
                candidate_dist = dist
        centroids.append(candidate)
    return centroids

# Returns the distance of a point from the closest point in points.
def min_dist(point, points):
    min_dist = euclidean_dist(point, points[0])
    for point2 in points:
        dist = euclidean_dist(point, point2)
        if dist < min_dist:
            min_dist = dist
    return min_dist

# Returns an Euclidean distance of two 2-dimensional points.
```

```
def euclidean_dist((x1, y1), (x2, y2)):
    return math.sqrt((x1 - x2) * (x1 - x2) + (y1 - y2) * (y1 - y2))

# PointGroup is a tuple that contains in the first coordinate a 2d point
# and in the second coordinate a group which a point is classified to.
def choose_centroids(point_groups, k):
    centroid_xs = [0] * k
    centroid_ys = [0] * k
    group_counts = [0] * k
    for ((x, y), group) in point_groups:
        centroid_xs[group] += x
        centroid_ys[group] += y
        group_counts[group] += 1
    centroids = []
    for group in range(0, k):
        centroids.append((
            float(centroid_xs[group]) / group_counts[group],
            float(centroid_ys[group]) / group_counts[group]))
    return centroids

# Returns the number of the centroid which is closest to the point.
# This number of the centroid is the number of the group where
# the point belongs to.
def closest_group(point, centroids):
    selected_group = 0
    selected_dist = euclidean_dist(point, centroids[0])
    for i in range(1, len(centroids)):
        dist = euclidean_dist(point, centroids[i])
        if dist < selected_dist:
            selected_group = i
            selected_dist = dist
    return selected_group

# Reassigns the groups to the points according to which centroid
# a point is closest to.
def assign_groups(point_groups, centroids):
    new_point_groups = []
    for (point, group) in point_groups:
        new_point_groups.append(
            (point, closest_group(point, centroids)))
    return new_point_groups

# Returns a list of pointgroups given a list of points.
def points_to_point_groups(points):
    point_groups = []
    for point in points:
        point_groups.append((point, 0))
    return point_groups
```

```
# Clusters points into the k groups adding every stage
# of the algorithm to the history which is returned.
def cluster_with_history(points, k):
    history = []
    centroids = choose_init_centroids(points, k)
    point_groups = points_to_point_groups(points)
    while True:
        point_groups = assign_groups(point_groups, centroids)
        history.append((point_groups, centroids))
        new_centroids = choose_centroids(point_groups, k)
        done = True
        for i in range(0, len(centroids)):
            if centroids[i] != new_centroids[i]:
                done = False
                break
        if done:
            return history
        centroids = new_centroids

# Program start
csv_file = sys.argv[1]
k = int(sys.argv[2])
everything = False
# The third argument sys.argv[3] represents the number of the step of the
# algorithm starting from 0 to be shown or "last" for displaying the last
# step and the number of the steps.
if sys.argv[3] == "last":
    everything = True
else:
    step = int(sys.argv[3])

data = common.csv_file_to_list(csv_file)
points = data_to_points(data)  # Represent every data item by a point.
history = cluster_with_history(points, k)
if everything:
    print "The total number of steps:", len(history)
    print "The history of the algorithm:"
    (point_groups, centroids) = history[len(history) - 1]
    # Print all the history.
    print_cluster_history(history)
    # But display the situation graphically at the last step only.
    draw(point_groups, centroids)
else:
    (point_groups, centroids) = history[step]
    print "Data for the step number", step, ":"
    print point_groups, centroids
    draw(point_groups, centroids)
```

Input data from gender classification

We save data from the gender classification example into the CSV file:

```
# source_code/5/persons_by_height_and_weight.csv
180,75
174,71
184,83
168,63
178,70
170,59
164,53
155,46
162,52
166,55
172,60
```

Program output for gender classification data

We run the program implementing k-means clustering algorithm on the data from the gender classification example. The numerical argument 2 means that we would like to cluster the data into 2 clusters:

```
$ python k-means_clustering.py persons_by_height_weight.csv 2 last
The total number of steps: 2
The history of the algorithm:
Step number 0: point_groups = [((180.0, 75.0), 0), ((174.0, 71.0), 0),
((184.0, 83.0), 0), ((168.0, 63.0), 0), ((178.0, 70.0), 0), ((170.0, 59.0),
0), ((164.0, 53.0), 1), ((155.0, 46.0), 1), ((162.0, 52.0), 1), ((166.0,
55.0), 1), ((172.0, 60.0), 0)]
centroids = [(180.0, 75.0), (155.0, 46.0)]
Step number 1: point_groups = [((180.0, 75.0), 0), ((174.0, 71.0), 0),
((184.0, 83.0), 0), ((168.0, 63.0), 0), ((178.0, 70.0), 0), ((170.0, 59.0),
0), ((164.0, 53.0), 1), ((155.0, 46.0), 1), ((162.0, 52.0), 1), ((166.0,
55.0), 1), ((172.0, 60.0), 0)]
centroids = [(175.14285714285714, 68.71428571428571), (161.75, 51.5)]
```

The program also outputs a graph visible in Image 5.2. The parameter `last` means that we would like the program to do the clustering until the last step. If we would like to display only the first step (step 0), we could change last to 0 to run:

```
$ python k-means_clustering.py persons_by_height_weight.csv 2 0
```

Upon the execution of the program, we would get the graph of the clusters and their centroids at the initial step as in Image 5.1.

House ownership – choosing the number of clusters

Let us take the example from the first chapter about the house ownership.

Age	Annual income in USD	House ownership status
23	50000	non-owner
37	34000	non-owner
48	40000	owner
52	30000	non-owner
28	95000	owner
25	78000	non-owner
35	130000	owner
32	105000	owner
20	100000	non-owner
40	60000	owner
50	80000	Peter

We would like to predict if Peter is a house owner using clustering.

Analysis:

Just as in the first chapter, we will have to scale the data since the income axis is by orders of magnitude greater and thus would diminish the impact of the age axis which actually has a good predictive power in this kind of problem. This is because it is expected that older people have had more time to settle down, save money and buy a house than the younger ones.

We apply the same rescaling from the Chapter 1 and get the following table:

Age	Scaled age	Annual income in USD	Scaled annual income	House ownership status
23	0.09375	50000	0.2	non-owner
37	0.53125	34000	0.04	non-owner
48	0.875	40000	0.1	owner

52	1	30000	0	non-owner
28	0.25	95000	0.65	owner
25	0.15625	78000	0.48	non-owner
35	0.46875	130000	1	owner
32	0.375	105000	0.75	owner
20	0	100000	0.7	non-owner
40	0.625	60000	0.3	owner
50	0.9375	80000	0.5	?

Given the table, we produce the input file for the algorithm and execute it, clustering the features into the two clusters.

Input:

```
# source_code/5/house_ownership2.csv
0.09375,0.2
0.53125,0.04
0.875,0.1
1,0
0.25,0.65
0.15625,0.48
0.46875,1
0.375,0.75
0,0.7
0.625,0.3
0.9375,0.5
```

Output for two clusters:

```
$ python k-means_clustering.py house_ownership2.csv 2 last
The total number of steps: 3
The history of the algorithm:
Step number 0: point_groups = [((0.09375, 0.2), 0), ((0.53125, 0.04), 0),
((0.875, 0.1), 1), ((1.0, 0.0), 1), ((0.25, 0.65), 0), ((0.15625, 0.48),
0), ((0.46875, 1.0), 0), ((0.375, 0.75), 0), ((0.0, 0.7), 0), ((0.625,
0.3), 1), ((0.9375, 0.5), 1)]
centroids = [(0.09375, 0.2), (1.0, 0.0)]
Step number 1: point_groups = [((0.09375, 0.2), 0), ((0.53125, 0.04), 1),
((0.875, 0.1), 1), ((1.0, 0.0), 1), ((0.25, 0.65), 0), ((0.15625, 0.48),
0), ((0.46875, 1.0), 0), ((0.375, 0.75), 0), ((0.0, 0.7), 0), ((0.625,
0.3), 1), ((0.9375, 0.5), 1)]
centroids = [(0.26785714285714285, 0.5457142857142857), (0.859375, 0.225)]
Step number 2: point_groups = [((0.09375, 0.2), 0), ((0.53125, 0.04), 1),
((0.875, 0.1), 1), ((1.0, 0.0), 1), ((0.25, 0.65), 0), ((0.15625, 0.48),
```

```
0), ((0.46875, 1.0), 0), ((0.375, 0.75), 0), ((0.0, 0.7), 0), ((0.625,
0.3), 1), ((0.9375, 0.5), 1)]
centroids = [(0.22395833333333334, 0.63), (0.79375, 0.188)]
```

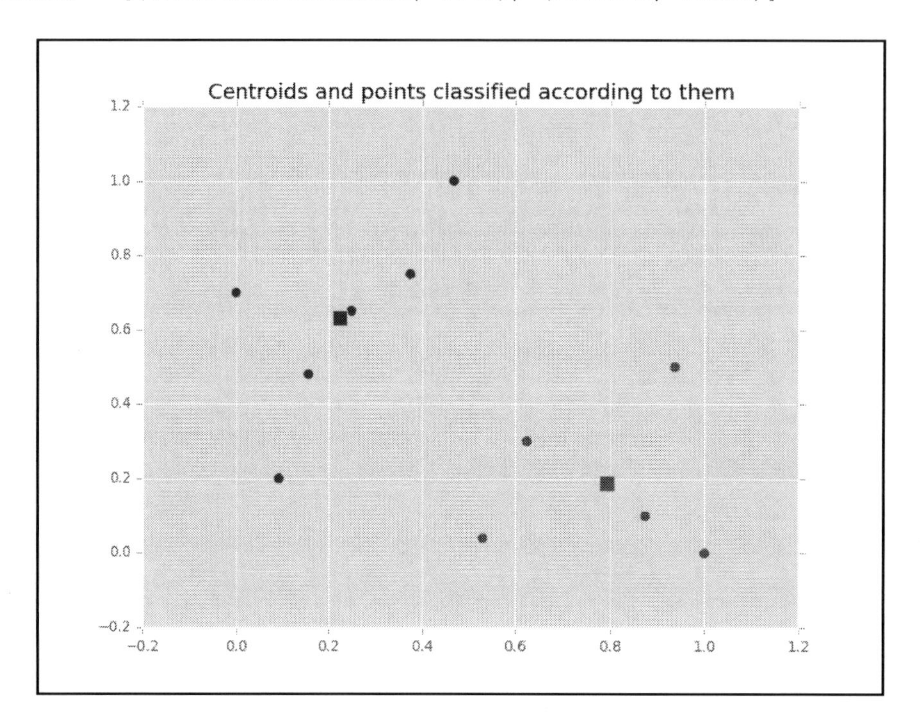

The blue cluster contains scaled features *(0.09375,0.2), (0.25,0.65), (0.15625,0.48), (0.46875,1), (0.375,0.75), (0,0.7)* or unscaled ones *(23,50000), (28,95000), (25,78000), (35,130000), (32,105000), (20,100000)*. The red cluster contains scaled features *(0.53125,0.04), (0.875,0.1), (1,0), (0.625,0.3), (0.9375,0.5)* or unscaled ones *(37,34000), (48,40000), (52,30000), (40,60000), (50,80000)*.

So Peter belongs to the red cluster. What is the proportion of house owners in a red cluster not counting Peter? 2/4 or 1/2 of the people in the red cluster are house owners. Thus the red cluster to which Peter belongs does not seem to have a high predictive power in determining whether Peter would be a house owner or not. We may try to cluster the data into more clusters in the hope that we would gain a purer cluster that could be more reliable for a prediction of the house-ownership for Peter. Let us therefore try to cluster the data into the three clusters.

Output for three clusters:

```
$ python k-means_clustering.py house_ownership2.csv 3 last
The total number of steps: 3
The history of the algorithm:
Step number 0: point_groups = [((0.09375, 0.2), 0), ((0.53125, 0.04), 0),
((0.875, 0.1), 1), ((1.0, 0.0), 1), ((0.25, 0.65), 2), ((0.15625, 0.48),
0), ((0.46875, 1.0), 2), ((0.375, 0.75), 2), ((0.0, 0.7), 0), ((0.625,
0.3), 1), ((0.9375, 0.5), 1)]
centroids = [(0.09375, 0.2), (1.0, 0.0), (0.46875, 1.0)]
Step number 1: point_groups = [((0.09375, 0.2), 0), ((0.53125, 0.04), 1),
((0.875, 0.1), 1), ((1.0, 0.0), 1), ((0.25, 0.65), 2), ((0.15625, 0.48),
0), ((0.46875, 1.0), 2), ((0.375, 0.75), 2), ((0.0, 0.7), 2), ((0.625,
0.3), 1), ((0.9375, 0.5), 1)]
centroids = [(0.1953125, 0.355), (0.859375, 0.225), (0.3645833333333333,
0.7999999999999999)]
Step number 2: point_groups = [((0.09375, 0.2), 0), ((0.53125, 0.04), 1),
((0.875, 0.1), 1), ((1.0, 0.0), 1), ((0.25, 0.65), 2), ((0.15625, 0.48),
0), ((0.46875, 1.0), 2), ((0.375, 0.75), 2), ((0.0, 0.7), 2), ((0.625,
0.3), 1), ((0.9375, 0.5), 1)]
centroids = [(0.125, 0.33999999999999997), (0.79375, 0.188), (0.2734375,
0.7749999999999999)]
```

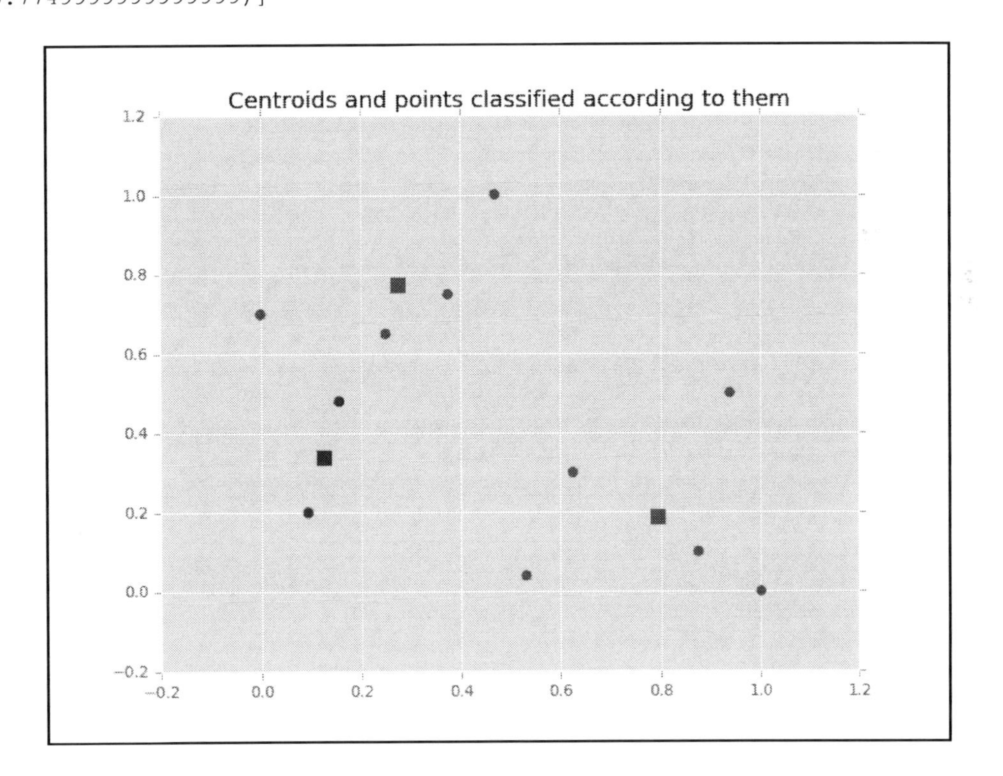

The red cluster has stayed the same. Let us therefore cluster the data into the 4 clusters.

Output for four clusters:

```
$ python k-means_clustering.py house_ownership2.csv 4 last
The total number of steps: 2
The history of the algorithm:
Step number 0: point_groups = [((0.09375, 0.2), 0), ((0.53125, 0.04), 0),
((0.875, 0.1), 1), ((1.0, 0.0), 1), ((0.25, 0.65), 3), ((0.15625, 0.48),
3), ((0.46875, 1.0), 2), ((0.375, 0.75), 2), ((0.0, 0.7), 3), ((0.625,
0.3), 1), ((0.9375, 0.5), 1)]
centroids = [(0.09375, 0.2), (1.0, 0.0), (0.46875, 1.0), (0.0, 0.7)]
Step number 1: point_groups = [((0.09375, 0.2), 0), ((0.53125, 0.04), 0),
((0.875, 0.1), 1), ((1.0, 0.0), 1), ((0.25, 0.65), 3), ((0.15625, 0.48),
3), ((0.46875, 1.0), 2), ((0.375, 0.75), 2), ((0.0, 0.7), 3), ((0.625,
0.3), 1), ((0.9375, 0.5), 1)]
centroids = [(0.3125, 0.12000000000000001), (0.859375, 0.225), (0.421875,
0.875), (0.13541666666666666, 0.61)]
```

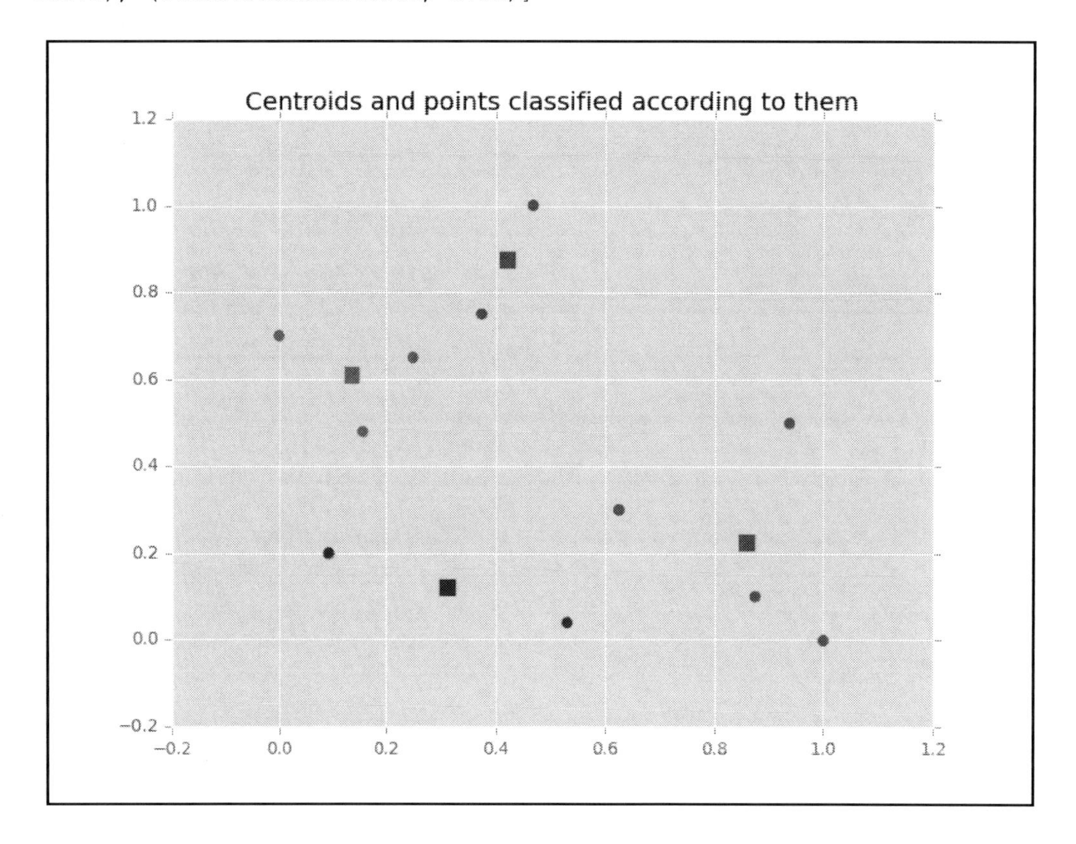

Now the red cluster where Peter belongs has changed. What is the proportion of the house owners in the red cluster now? If we do not count Peter, 2/3 of people in the red cluster own a house. When we clustered into the 2 or 3 clusters, the proportion was only ½ which did not tell us about the prediction of whether Peter is a house-owner or not. Now there is a majority of house owners in the red cluster not counting Peter, so we have a higher belief that Peter should also be a house owner. However, 2/3 is still a relatively low confidence for classifying Peter as a house owner. Let us partition the data into the 5 partitions to see what would happen.

Output for five clusters:

```
$ python k-means_clustering.py house_ownership2.csv 5 last
The total number of steps: 2
The history of the algorithm:
Step number 0: point_groups = [((0.09375, 0.2), 0), ((0.53125, 0.04), 0),
((0.875, 0.1), 1), ((1.0, 0.0), 1), ((0.25, 0.65), 3), ((0.15625, 0.48),
3), ((0.46875, 1.0), 2), ((0.375, 0.75), 2), ((0.0, 0.7), 3), ((0.625,
0.3), 4), ((0.9375, 0.5), 4)]
centroids = [(0.09375, 0.2), (1.0, 0.0), (0.46875, 1.0), (0.0, 0.7),
(0.9375, 0.5)]
Step number 1: point_groups = [((0.09375, 0.2), 0), ((0.53125, 0.04), 0),
((0.875, 0.1), 1), ((1.0, 0.0), 1), ((0.25, 0.65), 3), ((0.15625, 0.48),
3), ((0.46875, 1.0), 2), ((0.375, 0.75), 2), ((0.0, 0.7), 3), ((0.625,
0.3), 4), ((0.9375, 0.5), 4)]
centroids = [(0.3125, 0.12000000000000001), (0.9375, 0.05), (0.421875,
0.875), (0.13541666666666666, 0.61), (0.78125, 0.4)]
```

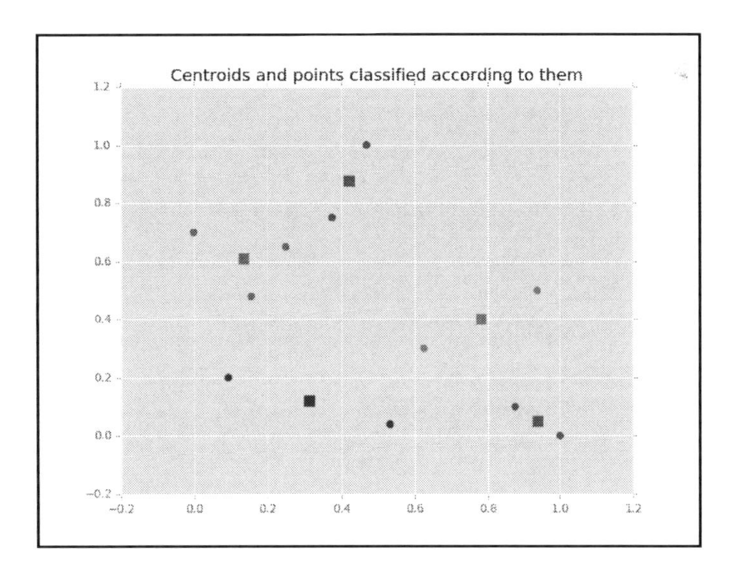

Now the red cluster contains only Peter and a non-owner. This clustering suggests that Peter is more likely a non-owner as well. However, according to the previous cluster Peter would be more likely an owner of a house. Therefore it may not be so clear whether Peter owns a house or not. Collecting more data would improve our analysis and should be carried out before making a definite classification in this problem.

From our analysis we noticed that a different number of clusters can result in a different result for a classification as the nature of members in an individual cluster can change. After collecting more data we should perform a cross-validation to determine the number of the clusters that classifies the data with the highest accuracy.

Document clustering – understanding the number of clusters k in a semantic context

We are given the following information about the frequency counts for the words money and god(s) in the following 17 books from the Project Gutenberg:

Book number	**Book name**	**Money in %**	**God(s) in %**
1	The Vedanta-Sutras with the Commentary by Ramanuja, by Trans. George Thibaut	0	0.07
2	The Mahabharata of Krishna-Dwaipayana Vyasa - Adi Parva, by Kisari Mohan Ganguli	0	0.17
3	The Mahabharata of Krishna-Dwaipayana Vyasa, Part 2, by Krishna-Dwaipayana Vyasa	0.01	0.10
4	Mahabharata of Krishna-Dwaipayana Vyasa Bk. 3 Pt. 1, by Krishna-Dwaipayana Vyasa	0	0.32
5	The Mahabharata of Krishna-Dwaipayana Vyasa Bk. 4, by Kisari Mohan Ganguli	0	0.06
6	The Mahabharata of Krishna-Dwaipayana Vyasa Bk. 3 Pt. 2, by Translated by Kisari Mohan Ganguli	0	0.27
7	The Vedanta-Sutras with the Commentary by Sankaracarya	0	0.06
8	The King James Bible	0.02	0.59

9	Paradise Regained, by John Milton	0.02	0.45
10	Imitation of Christ, by Thomas A Kempis	0.01	0.69
11	The Koran as translated by Rodwell	0.01	1.72
12	The Adventures of Tom Sawyer, Complete by Mark Twain (Samuel Clemens)	0.05	0.01
13	Adventures of Huckleberry Finn, Complete by Mark Twain (Samuel Clemens)	0.08	0
14	Great Expectations, by Charles Dickens	0.04	0.01
15	The Picture of Dorian Gray, by Oscar Wilde	0.03	0.03
16	The Adventures of Sherlock Holmes, by Arthur Conan Doyle	0.04	0.03
17	Metamorphosis, by Franz Kafka Translated by David Wyllie	0.06	0.03

We would like to cluster this dataset based on the on the chosen frequency counts of the words into the groups by their semantic context.

Analysis:

First we will do a rescaling since the highest frequency count of the word money is 0.08% whereas the highest frequency count of the word god(s) is 1.72%. So we will divide the frequency counts of money by 0.08 and the frequency counts of god(s) by 1.72:

Book number	Money scaled	God(s) scaled
1	0	0.0406976744
2	0	0.0988372093
3	0.125	0.0581395349
4	0	0.1860465116
5	0	0.0348837209
6	0	0.1569767442
7	0	0.0348837209
8	0.25	0.3430232558
9	0.25	0.261627907

10	0.125	0.4011627907
11	0.125	1
12	0.625	0.0058139535
13	1	0
14	0.5	0.0058139535
15	0.375	0.0174418605
16	0.5	0.0174418605
17	0.75	0.0174418605

Now that we have rescaled data, let us apply k-means clustering algorithm trying dividing the data into a different number of the clusters.

Input:

```
source_code/5/document_clustering/word_frequencies_money_god_scaled.csv
0,0.0406976744
0,0.0988372093
0.125,0.0581395349
0,0.1860465116
0,0.0348837209
0,0.1569767442
0,0.0348837209
0.25,0.3430232558
0.25,0.261627907
0.125,0.4011627907
0.125,1
0.625,0.0058139535
1,0
0.5,0.0058139535
0.375,0.0174418605
0.5,0.0174418605
0.75,0.0174418605
```

Output for 2 clusters:

```
$ python k-means_clustering.py
document_clustering/word_frequencies_money_god_scaled.csv 2 last
The total number of steps: 3
The history of the algorithm:
Step number 0: point_groups = [((0.0, 0.0406976744), 0), ((0.0,
0.0988372093), 0), ((0.125, 0.0581395349), 0), ((0.0, 0.1860465116), 0),
((0.0, 0.0348837209), 0), ((0.0, 0.1569767442), 0), ((0.0, 0.0348837209),
0), ((0.25, 0.3430232558), 0), ((0.25, 0.261627907), 0), ((0.125,
```

```
0.4011627907), 0), ((0.125, 1.0), 0), ((0.625, 0.0058139535), 1), ((1.0,
0.0), 1), ((0.5, 0.0058139535), 1), ((0.375, 0.0174418605), 0), ((0.5,
0.0174418605), 1), ((0.75, 0.0174418605), 1)]
centroids = [(0.0, 0.0406976744), (1.0, 0.0)]
Step number 1: point_groups = [((0.0, 0.0406976744), 0), ((0.0,
0.0988372093), 0), ((0.125, 0.0581395349), 0), ((0.0, 0.1860465116), 0),
((0.0, 0.0348837209), 0), ((0.0, 0.1569767442), 0), ((0.0, 0.0348837209),
0), ((0.25, 0.3430232558), 0), ((0.25, 0.261627907), 0), ((0.125,
0.4011627907), 0), ((0.125, 1.0), 0), ((0.625, 0.0058139535), 1), ((1.0,
0.0), 1), ((0.5, 0.0058139535), 1), ((0.375, 0.0174418605), 1), ((0.5,
0.0174418605), 1), ((0.75, 0.0174418605), 1)]
centroids = [(0.10416666666666667, 0.21947674418333332), (0.675,
0.0093023256)]
Step number 2: point_groups = [((0.0, 0.0406976744), 0), ((0.0,
0.0988372093), 0), ((0.125, 0.0581395349), 0), ((0.0, 0.1860465116), 0),
((0.0, 0.0348837209), 0), ((0.0, 0.1569767442), 0), ((0.0, 0.0348837209),
0), ((0.25, 0.3430232558), 0), ((0.25, 0.261627907), 0), ((0.125,
0.4011627907), 0), ((0.125, 1.0), 0), ((0.625, 0.0058139535), 1), ((1.0,
0.0), 1), ((0.5, 0.0058139535), 1), ((0.375, 0.0174418605), 1), ((0.5,
0.0174418605), 1), ((0.75, 0.0174418605), 1)]
centroids = [(0.07954545454545454, 0.2378435517909091), (0.625,
0.01065891475)]
```

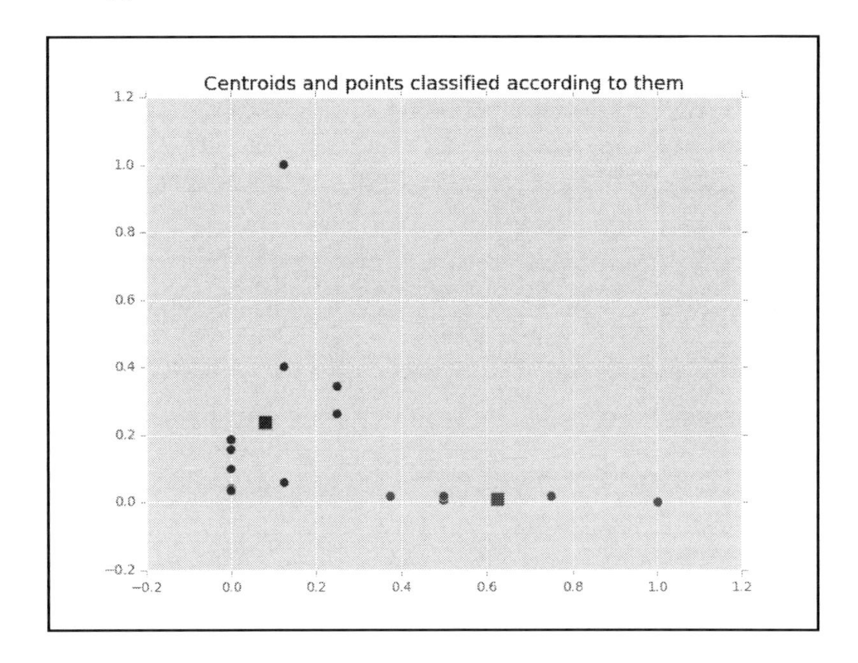

We can observe that clustering into the 2 clusters divides books into religious ones, the ones in the blue cluster and non-religious ones, the ones in the red cluster. Let us try to cluster the books into the 3 clusters to observe how the algorithm would divide the data.

Output for 3 clusters:

```
$ python k-means_clustering.py
document_clustering/word_frequencies_money_god_scaled.csv 3 last
The total number of steps: 3
The history of the algorithm:
Step number 0: point_groups = [((0.0, 0.0406976744), 0), ((0.0,
0.0988372093), 0), ((0.125, 0.0581395349), 0), ((0.0, 0.1860465116), 0),
((0.0, 0.0348837209), 0), ((0.0, 0.1569767442), 0), ((0.0, 0.0348837209),
0), ((0.25, 0.3430232558), 0), ((0.25, 0.261627907), 0), ((0.125,
0.4011627907), 0), ((0.125, 1.0), 2), ((0.625, 0.0058139535), 1), ((1.0,
0.0), 1), ((0.5, 0.0058139535), 1), ((0.375, 0.0174418605), 0), ((0.5,
0.0174418605), 1), ((0.75, 0.0174418605), 1)]
centroids = [(0.0, 0.0406976744), (1.0, 0.0), (0.125, 1.0)]
Step number 1: point_groups = [((0.0, 0.0406976744), 0), ((0.0,
0.0988372093), 0), ((0.125, 0.0581395349), 0), ((0.0, 0.1860465116), 0),
((0.0, 0.0348837209), 0), ((0.0, 0.1569767442), 0), ((0.0, 0.0348837209),
0), ((0.25, 0.3430232558), 0), ((0.25, 0.261627907), 0), ((0.125,
0.4011627907), 0), ((0.125, 1.0), 2), ((0.625, 0.0058139535), 1), ((1.0,
0.0), 1), ((0.5, 0.0058139535), 1), ((0.375, 0.0174418605), 1), ((0.5,
0.0174418605), 1), ((0.75, 0.0174418605), 1)]
centroids = [(0.10227272727272728, 0.14852008456363636), (0.675,
0.0093023256), (0.125, 1.0)]
Step number 2: point_groups = [((0.0, 0.0406976744), 0), ((0.0,
0.0988372093), 0), ((0.125, 0.0581395349), 0), ((0.0, 0.1860465116), 0),
((0.0, 0.0348837209), 0), ((0.0, 0.1569767442), 0), ((0.0, 0.0348837209),
0), ((0.25, 0.3430232558), 0), ((0.25, 0.261627907), 0), ((0.125,
0.4011627907), 0), ((0.125, 1.0), 2), ((0.625, 0.0058139535), 1), ((1.0,
0.0), 1), ((0.5, 0.0058139535), 1), ((0.375, 0.0174418605), 1), ((0.5,
0.0174418605), 1), ((0.75, 0.0174418605), 1)]
centroids = [(0.075, 0.16162790697), (0.625, 0.01065891475), (0.125, 1.0)]
```

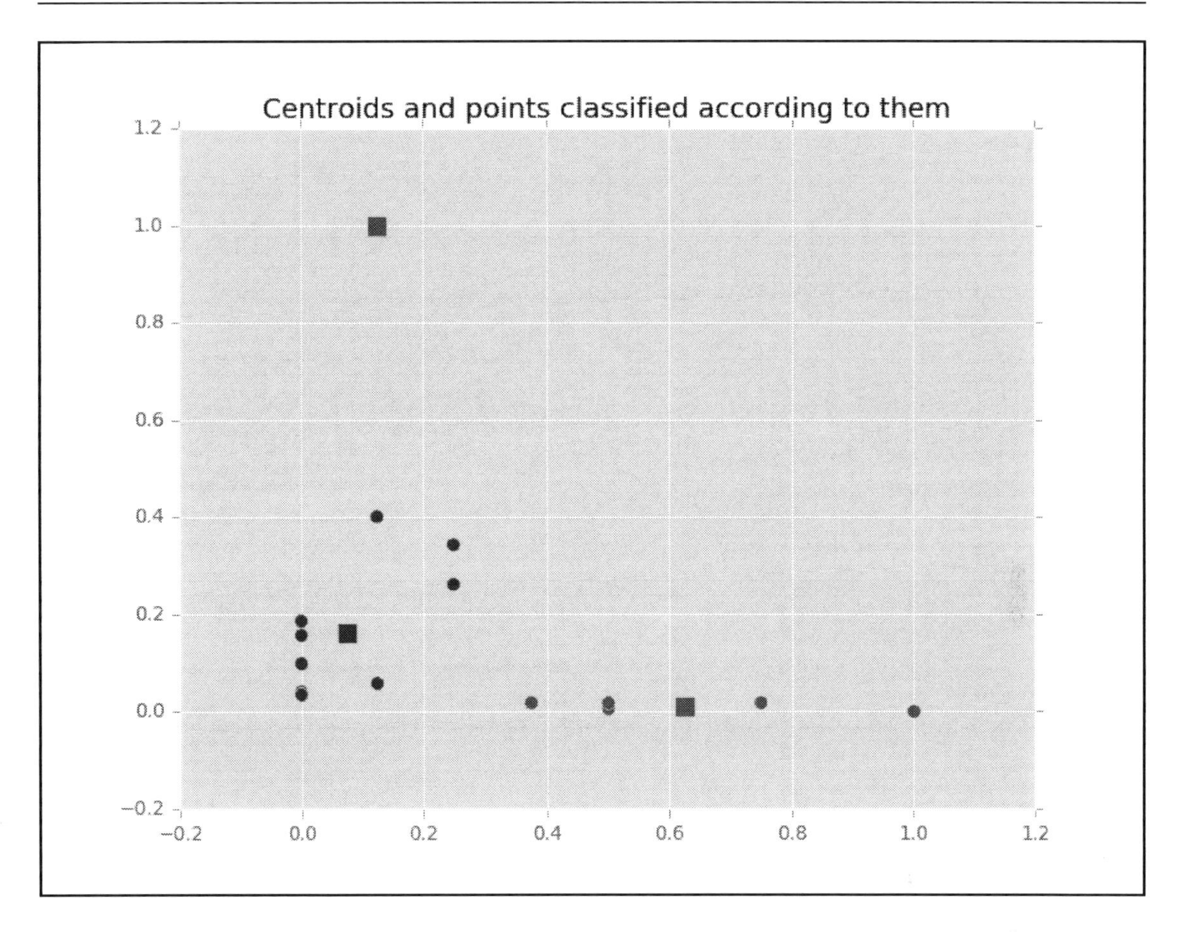

This time the algorithm separated from the religious books book The Koran into a green cluster. This is because in fact the word god is the 5th most frequent word in The Koran. The clustering here happens to divide the books according to the writing style they were written with. Clustering into 4 clusters separates one book that has a relatively high frequency of the word money from the red cluster of non-religious books into a separate cluster. Let us look at the clustering into the 5 clusters.

Output for 5 clusters:

```
$ python k-means_clustering.py word_frequencies_money_god_scaled.csv 5 last
The total number of steps: 2
The history of the algorithm:
Step number 0: point_groups = [((0.0, 0.0406976744), 0), ((0.0,
0.0988372093), 0), ((0.125, 0.0581395349), 0), ((0.0, 0.1860465116), 0),
((0.0, 0.0348837209), 0), ((0.0, 0.1569767442), 0), ((0.0, 0.0348837209),
0), ((0.25, 0.3430232558), 4), ((0.25, 0.261627907), 4), ((0.125,
```

```
0.4011627907), 4), ((0.125, 1.0), 2), ((0.625, 0.0058139535), 3), ((1.0,
0.0), 1), ((0.5, 0.0058139535), 3), ((0.375, 0.0174418605), 3), ((0.5,
0.0174418605), 3), ((0.75, 0.0174418605), 3)]
centroids = [(0.0, 0.0406976744), (1.0, 0.0), (0.125, 1.0), (0.5,
0.0174418605), (0.25, 0.3430232558)]
Step number 1: point_groups = [((0.0, 0.0406976744), 0), ((0.0,
0.0988372093), 0), ((0.125, 0.0581395349), 0), ((0.0, 0.1860465116), 0),
((0.0, 0.0348837209), 0), ((0.0, 0.1569767442), 0), ((0.0, 0.0348837209),
0), ((0.25, 0.3430232558), 4), ((0.25, 0.261627907), 4), ((0.125,
0.4011627907), 4), ((0.125, 1.0), 2), ((0.625, 0.0058139535), 3), ((1.0,
0.0), 1), ((0.5, 0.0058139535), 3), ((0.375, 0.0174418605), 3), ((0.5,
0.0174418605), 3), ((0.75, 0.0174418605), 3)]
centroids = [(0.017857142857142856, 0.08720930231428571), (1.0, 0.0),
(0.125, 1.0), (0.55, 0.0127906977), (0.20833333333333334,
0.3352713178333333)]
```

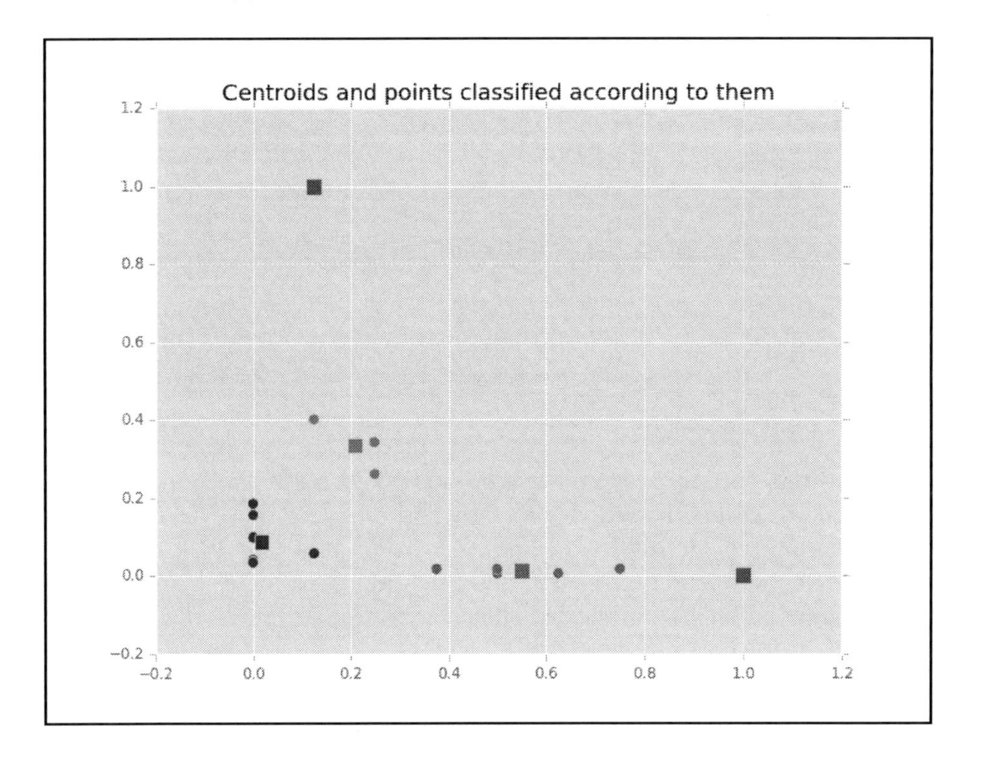

This clustering further divides the blue cluster of the remaining religious books into the blue cluster of the Hindi books and the gray cluster of the Christian books.

We can use clustering this way to group items with similar properties and then enable to find similar items quickly based on the given example. The granularity of the clustering parameter k determines how similar we can expect the items in a group to be. The higher the parameter, the more similar items are going to be in the cluster, but a smaller number of them.

Summary

Clustering of the data is very efficient and can be used to facilitate a faster classification of the new features by classifying a feature to the class represented in the cluster of that feature. An appropriate number of the clusters can be determined by cross-validation choosing the one that results in the most accurate classification.

Clustering orders data by their similarity. The more clusters, the greater similarity between the features in a cluster, but a fewer features in a cluster.

The k-means clustering algorithm is a clustering algorithm that tries to cluster features in such a way that the mutual distance of the features in a cluster is minimized. To do this, the algorithm computes centroid of each cluster and a feature belongs to the cluster whose centroid is closest to it. The algorithm finishes the computation of the clusters as soon as they or their centroids no longer change.

Problems

1. Compute the centroid of the following clusters:
    1. 2, 3, 4
    2. 100$, 400$, 1000$
    3. (10,20), (40, 60), (0, 40)
    4. (200$, 40km), (300$, 60km), (500$, 100km), (250$, 200km)
    5. (1,2,4), (0,0,3), (10,20,5), (4,8,2), (5,0,1)
2. Cluster the following datasets into the 2, 3 and 4 clusters using k-means clustering algorithm:
    1. 0, 2, 5, 4, 8, 10, 12, 11.
    2. (2,2), (2,5), (10,4), (3,5), (7,3), (5,9), (2,8), (4,10), (7,4), (4,4), (5,8), (9,3).

3. (Couples and the number of their children) We are given the following ages of the couples and their number of the children.

Couple number	Wife age	Husband age	Number of children
1	48	49	5
2	40	43	2
3	24	28	1
4	49	42	3
5	32	34	0
6	24	27	0
7	29	32	2
8	35	35	2
9	33	36	1
10	42	47	3
11	22	27	2
12	41	45	4
13	39	43	4
14	36	38	2
15	30	32	1
16	36	38	0
17	36	39	3
18	37	38	?

We would like to guess using clustering how many children a couple has where the age of the husband is 37 and the age of the wife is 38.

Analysis:

1. (1/3)*(2+3+4)=3
    1. (1/3)*(100$+400$+1000$)=500$
    2. ((10+40+0)/3,(20+60+40)/3)=(50/3, 120/3)=(50/3, 40)
    3. ((200$+300$+500$+250$)/4,(40km+60km+100km+200km)/4)
    4. =(1250$/4,400km/4)=(312.5$,100km)
    5. ((1+0+10+4+5)/5,(2+0+20+8+0)/5,(4+3+5+2+1)/5)=(4,6,3)
2. We add a second coordinate and set it to 0 for all the features. This way the distance between the features does not change and we can use the clustering algorithm we implemented earlier in this chapter.

   Input:

```
# source_code/5/problem5_2.csv
0,0
2,0
5,0
4,0
8,0
10,0
12,0
11,0
```

For 2 clusters:

```
$ python k-means_clustering.py problem5_2.csv 2 last
The total number of steps: 2
The history of the algorithm:
Step number 0: point_groups = [((0.0, 0.0), 0), ((2.0, 0.0),
0), ((5.0, 0.0), 0), ((4.0, 0.0), 0), ((8.0, 0.0), 1), ((10.0,
0.0), 1), ((12.0, 0.0), 1), ((11.0, 0.0), 1)]
centroids = [(0.0, 0.0), (12.0, 0.0)]
Step number 1: point_groups = [((0.0, 0.0), 0), ((2.0, 0.0),
0), ((5.0, 0.0), 0), ((4.0, 0.0), 0), ((8.0, 0.0), 1), ((10.0,
0.0), 1), ((12.0, 0.0), 1), ((11.0, 0.0), 1)]
centroids = [(2.75, 0.0), (10.25, 0.0)]
```

For 3 clusters:

```
$ python k-means_clustering.py problem5_2.csv 3 last
The total number of steps: 2
The history of the algorithm:
Step number 0: point_groups = [((0.0, 0.0), 0), ((2.0, 0.0),
0), ((5.0, 0.0), 2), ((4.0, 0.0), 2), ((8.0, 0.0), 2), ((10.0,
0.0), 1), ((12.0, 0.0), 1), ((11.0, 0.0), 1)]
```

```
centroids = [(0.0, 0.0), (12.0, 0.0), (5.0, 0.0)]
Step number 1: point_groups = [((0.0, 0.0), 0), ((2.0, 0.0),
0), ((5.0, 0.0), 2), ((4.0, 0.0), 2), ((8.0, 0.0), 2), ((10.0,
0.0), 1), ((12.0, 0.0), 1), ((11.0, 0.0), 1)]
centroids = [(1.0, 0.0), (11.0, 0.0), (5.666666666666667, 0.0)]
```

For 4 clusters:

```
$ python k-means_clustering.py problem5_2.csv 4 last
The total number of steps: 2
The history of the algorithm:
Step number 0: point_groups = [((0.0, 0.0), 0), ((2.0, 0.0),
0), ((5.0, 0.0), 2), ((4.0, 0.0), 2), ((8.0, 0.0), 3), ((10.0,
0.0), 1), ((12.0, 0.0), 1), ((11.0, 0.0), 1)]
centroids = [(0.0, 0.0), (12.0, 0.0), (5.0, 0.0), (8.0, 0.0)]
Step number 1: point_groups = [((0.0, 0.0), 0), ((2.0, 0.0),
0), ((5.0, 0.0), 2), ((4.0, 0.0), 2), ((8.0, 0.0), 3), ((10.0,
0.0), 1), ((12.0, 0.0), 1), ((11.0, 0.0), 1)]
centroids = [(1.0, 0.0), (11.0, 0.0), (4.5, 0.0), (8.0, 0.0)]
```

b) We use the implemented algorithm again.

Input:

```
# source_code/5/problem5_2b.csv
2,2
2,5
10,4
3,5
7,3
5,9
2,8
4,10
7,4
4,4
5,8
9,3
```

Output for 2 clusters:

```
$ python k-means_clustering.py problem5_2b.csv 2 last
The total number of steps: 3
The history of the algorithm:
Step number 0: point_groups = [((2.0, 2.0), 0), ((2.0, 5.0),
0), ((10.0, 4.0), 1), ((3.0, 5.0), 0), ((7.0, 3.0), 1), ((5.0,
9.0), 1), ((2.0, 8.0), 0), ((4.0, 10.0), 0), ((7.0, 4.0), 1),
((4.0, 4.0), 0), ((5.0, 8.0), 1), ((9.0, 3.0), 1)]
centroids = [(2.0, 2.0), (10.0, 4.0)]
Step number 1: point_groups = [((2.0, 2.0), 0), ((2.0, 5.0),
```

```
0), ((10.0, 4.0), 1), ((3.0, 5.0), 0), ((7.0, 3.0), 1), ((5.0,
9.0), 0), ((2.0, 8.0), 0), ((4.0, 10.0), 0), ((7.0, 4.0), 1),
((4.0, 4.0), 0), ((5.0, 8.0), 0), ((9.0, 3.0), 1)]
centroids = [(2.8333333333333335, 5.666666666666667),
(7.166666666666667, 5.166666666666667)]
Step number 2: point_groups = [((2.0, 2.0), 0), ((2.0, 5.0),
0), ((10.0, 4.0), 1), ((3.0, 5.0), 0), ((7.0, 3.0), 1), ((5.0,
9.0), 0), ((2.0, 8.0), 0), ((4.0, 10.0), 0), ((7.0, 4.0), 1),
((4.0, 4.0), 0), ((5.0, 8.0), 0), ((9.0, 3.0), 1)]
centroids = [(3.375, 6.375), (8.25, 3.5)]
```

Output for 3 clusters:

```
$ python k-means_clustering.py problem5_2b.csv 3 last
The total number of steps: 2
The history of the algorithm:
Step number 0: point_groups = [((2.0, 2.0), 0), ((2.0, 5.0),
0), ((10.0, 4.0), 1), ((3.0, 5.0), 0), ((7.0, 3.0), 1), ((5.0,
9.0), 2), ((2.0, 8.0), 2), ((4.0, 10.0), 2), ((7.0, 4.0), 1),
((4.0, 4.0), 0), ((5.0, 8.0), 2), ((9.0, 3.0), 1)]
centroids = [(2.0, 2.0), (10.0, 4.0), (4.0, 10.0)]
Step number 1: point_groups = [((2.0, 2.0), 0), ((2.0, 5.0),
0), ((10.0, 4.0), 1), ((3.0, 5.0), 0), ((7.0, 3.0), 1), ((5.0,
9.0), 2), ((2.0, 8.0), 2), ((4.0, 10.0), 2), ((7.0, 4.0), 1),
((4.0, 4.0), 0), ((5.0, 8.0), 2), ((9.0, 3.0), 1)]
centroids = [(2.75, 4.0), (8.25, 3.5), (4.0, 8.75)]
```

Output for 4 clusters:

```
$ python k-means_clustering.py problem5_2b.csv 4 last
The total number of steps: 2
The history of the algorithm:
Step number 0: point_groups = [((2.0, 2.0), 0), ((2.0, 5.0),
3), ((10.0, 4.0), 1), ((3.0, 5.0), 3), ((7.0, 3.0), 1), ((5.0,
9.0), 2), ((2.0, 8.0), 2), ((4.0, 10.0), 2), ((7.0, 4.0), 1),
((4.0, 4.0), 3), ((5.0, 8.0), 2), ((9.0, 3.0), 1)]
centroids = [(2.0, 2.0), (10.0, 4.0), (4.0, 10.0), (3.0, 5.0)]
Step number 1: point_groups = [((2.0, 2.0), 0), ((2.0, 5.0),
3), ((10.0, 4.0), 1), ((3.0, 5.0), 3), ((7.0, 3.0), 1), ((5.0,
9.0), 2), ((2.0, 8.0), 2), ((4.0, 10.0), 2), ((7.0, 4.0), 1),
((4.0, 4.0), 3), ((5.0, 8.0), 2), ((9.0, 3.0), 1)]
centroids = [(2.0, 2.0), (8.25, 3.5), (4.0, 8.75), (3.0,
4.666666666666667)]
```

3. We are given 17 couples and their number of children and would like to find out how many children has the 18th couple. We will use the first 14 couples as data and then the next 3 couples for the cross-validation to determine the number of clusters k that we will use to find out how many children the 18th couple is expected to have.

   After clustering we will say that a couple is likely to have about the number of the children that is the average of the children in that cluster. Using the cross-validation we will choose the number of the clusters that will minimize the difference between the actual number of the children and the predicted number of the children. We will capture this difference for all the items in the cluster cumulatively as the square root of the squares of the differences of children for each couple. This will minimize the variance of the random variable for the predicted number of the children for the 18th couple.

   We will perform the clustering into 2,3,4 and 5 clusters.

   Input:

```
# source_code/5/couples_children.csv
48,49
40,43
24,28
49,42
32,34
24,27
29,32
35,35
33,36
42,47
22,27
41,45
39,43
36,38
30,32
36,38
36,39
37,38
```

Output for 2 clusters:

A couple listed for a cluster is of the form (couple_number,(wife_age,husband_age)).

```
Cluster 0: [(1, (48.0, 49.0)), (2, (40.0, 43.0)), (4, (49.0,
42.0)), (10, (42.0, 47.0)), (12, (41.0, 45.0)), (13, (39.0,
```

```
43.0)), (14, (36.0, 38.0)), (16, (36.0, 38.0)), (17, (36.0,
39.0)), (18, (37.0, 38.0))]
Cluster 1: [(3, (24.0, 28.0)), (5, (32.0, 34.0)), (6, (24.0,
27.0)), (7, (29.0, 32.0)), (8, (35.0, 35.0)), (9, (33.0,
36.0)), (11, (22.0, 27.0)), (15, (30.0, 32.0))]
```

We would like to determine the expected number of the children for the 15th couple *(30,32)*, i.e where a wife is 30 years old and the husband is 32 years old. *(30,32)* is in the cluster 1. The couples in the cluster 1 are: *(24.0, 28.0), (32.0, 34.0), (24.0, 27.0), (29.0, 32.0), (35.0, 35.0), (33.0, 36.0), (22.0, 27.0), (30.0, 32.0)*. Out of these and the first 14 couples used for the data the remaining couples are: *(24.0, 28.0), (32.0, 34.0), (24.0, 27.0), (29.0, 32.0), (35.0, 35.0), (33.0, 36.0), (22.0, 27.0)*. The average number of the children for these couples is *est15=8/7~1.14*. This is the estimated number of the children for the 15th couple based on the data from the first 14 couples.

The estimated number of the children for the 16th couple is est16=23/7~3.29. The estimated number of the children for the 17th couple is also *est17=23/7~3.29* since both 16th and 17th couple belong to the same cluster.

Now we will calculate the error E2 (2 for 2 clusters) between the estimated number of the children (e.g. denoted est15 for the 15th couple) and the actual number of the children (example. denoted act15 for the 15th couple) as follows:

E2=sqrt(sqr(est15-act15)+sqr(est16-act16)+sqr(est17-act17))

=sqrt(sqr(8/7-1)+sqr(23/7-0)+sqr(23/7-3))~3.3

Now that we have calculated the error E2, we will calculate the errors of the estimation with the other number of clusters. We will choose the number of the clusters with the least error to estimate the number of the children for the 18th couple.

Output for 3 clusters:

```
Cluster 0: [(1, (48.0, 49.0)), (2, (40.0, 43.0)), (4, (49.0, 42.0)), (10,
(42.0, 47.0)), (12, (41.0, 45.0)), (13, (39.0, 43.0))]
Cluster 1: [(3, (24.0, 28.0)), (6, (24.0, 27.0)), (7, (29.0, 32.0)), (11,
(22.0, 27.0)), (15, (30.0, 32.0))]
Cluster 2: [(5, (32.0, 34.0)), (8, (35.0, 35.0)), (9, (33.0, 36.0)), (14,
(36.0, 38.0)), (16, (36.0, 38.0)), (17, (36.0, 39.0)), (18, (37.0, 38.0))]
```

Now the 15th couple is in the cluster 1, 16th couple in the cluster 2, 17th couple in the cluster 2. So the estimated number of the children for each couple is *5/4=1.25*.

The error E3 of the estimation is:

$E3=sqrt((1.25-1)^2+(1.25-0)^2+(1.25-3)^2)\sim 2.17$

Output for 4 clusters:

```
Cluster 0: [(1, (48.0, 49.0)), (4, (49.0, 42.0)), (10, (42.0, 47.0)), (12,
(41.0, 45.0))]
Cluster 1: [(3, (24.0, 28.0)), (6, (24.0, 27.0)), (11, (22.0, 27.0))]
Cluster 2: [(2, (40.0, 43.0)), (13, (39.0, 43.0)), (14, (36.0, 38.0)), (16,
(36.0, 38.0)), (17, (36.0, 39.0)), (18, (37.0, 38.0))]
Cluster 3: [(5, (32.0, 34.0)), (7, (29.0, 32.0)), (8, (35.0, 35.0)), (9,
(33.0, 36.0)), (15, (30.0, 32.0))]
```

The 15th couple is in the cluster 3, 16th in the cluster 2, 17th in the cluster 2. So the estimated number of the children for the 15th couple is *5/4=1.25*. The estimated number of the children for the 16th and 17th couple is 8/3~2.67 children.

The error E4 of the estimation is:

$E4=sqrt((1.25-1)^2+(8/3-0)^2+(8/3-3)^2)\sim 2.70$

Output for 5 clusters:

```
Cluster 0: [(1, (48.0, 49.0)), (4, (49.0, 42.0))]
Cluster 1: [(3, (24.0, 28.0)), (6, (24.0, 27.0)), (11, (22.0, 27.0))]
Cluster 2: [(8, (35.0, 35.0)), (9, (33.0, 36.0)), (14, (36.0, 38.0)), (16,
(36.0, 38.0)), (17, (36.0, 39.0)), (18, (37.0, 38.0))]
Cluster 3: [(5, (32.0, 34.0)), (7, (29.0, 32.0)), (15, (30.0, 32.0))]
Cluster 4: [(2, (40.0, 43.0)), (10, (42.0, 47.0)), (12, (41.0, 45.0)), (13,
(39.0, 43.0))]
```

The 15th couple is in the cluster 3, 16th in the cluster 2, 17th in the cluster 2. So the estimated number of the children for the 15th couple is 1. The estimated number of the children for the 16th and 17th couple is 5/3~1.67.

The error E5 of the estimation is:

$E5=sqrt((1-1)^2+(5/3-0)^2+(5/3-3)^2)\sim 2.13$

Using cross-validation to determine the outcome:

We used 14 couples as data for the estimation and 3 other couples for cross-validation to find the best parameter of k clusters among the values 2,3,4,5. We may try to cluster into more clusters, but since we have so relatively very little data, it should be sufficient to cluster into the 5 clusters at most. Let us summarize the errors of the estimation.

Number of clusters	Error rate
2	3.3
3	2.17
4	2.7
5	2.13

The error rate is the least for 3 and 5 clusters. The fact that the error rate goes up for 4 clusters and then down again for 5 clusters may indicate that we may not have enough data to make a good estimate. A natural expectation would be that there are not local maxims of errors for the values of k greater than 2. Moreover the difference between the error for clustering with 3 and 5 clusters is very small and one cluster out of 5 is smaller than one cluster out of 3. For this reason we choose 3 clusters over 5 to estimate the number of the children for the 18th couple.

When clustering into the 3 clusters, 18th couple is in the cluster 2. Therefore the estimated number of the children for the 18th couple is 1.25.

6
Regression

Regression analysis is a process of estimating the relationship between dependent variables. For example, if a variable y is linearly dependent on the variable x, then regression analysis tries to estimate the constants a and b in the equation $y=ax+b$ that expresses the linear relationship between the variables y and x.

In this chapter, you will learn the following:

- The core idea of a regression by performing a simple linear regression on the perfect data from the first principles in example Fahrenheit and Celsius conversion
- Linear regression analysis in the statistical software R on perfect and real-world data in examples Fahrenheit and Celsius conversion, weight prediction from height, and flight time duration prediction from the distance
- The gradient descent algorithm to find a regression model with the best fit (using least mean squares rule) and how to implement it in Python in section *Gradient descent algorithm and its implementation*
- How to find a non-linear regression model using R in example ballistic flight analysis and problem 4, bacteria population prediction

Fahrenheit and Celsius conversion - linear regression on perfect data

For example, Fahrenheit and Celsius degrees are related in a linear way. Given a table with pairs of both Fahrenheit and Celsius degrees, we can estimate the constants to devise a conversion formula from degrees Fahrenheit to degrees Celsius or vice versa:

^{0}F	^{0}C
5	-15
14	-10
23	-5
32	0
41	5
50	10

Analysis from first principles:

We would like to derive a formula converting *F* (degrees Fahrenheit) to *C* (degrees Celsius) as follows:

$$C=a*F+b$$

Here, a and b are the constants to be found. A graph of the function $C=a*F+b$ is a straight line and thus is uniquely determined by two points. Therefore, we actually need only the two points from the table, say pairs *(F1,C1)* and *(F2,C2)*. Then we will have the following:

$$C1=a*F1+b \quad C2=a*F2+b$$

Now, $C2-C1=(a*F2+b)-(a*F1+b)=a*(F2-F1)$. Therefore, we have the following:

$$a=(C2-C1)/(F2-F1)$$

$$b=C1-a*F1=C1-[(C2-C1)/(F2-F1)]$$

So let us take for example the first two pairs *(F1,C1)=(5,-15)* and *(F2,C2)=(14,-10)*, then we have the following:

$$a=(-10-(-15))/(14-5)=5/9$$

$$b=-15-(5/9)*5=-160/9$$

Therefore, the formula to calculate degrees Celsius from degrees Fahrenheit is *C=(5/9)*F-160/9~0.5556*F-17.7778*.

Let us verify it against the data in the table:

^{0}F	^{0}C	(5/9)*F-160/9
5	-15	-15
14	-10	-10
23	-5	-5
32	0	0
41	5	5
50	10	10

Therefore, the formula fits our input data 100%. The data we worked with was perfect. In later examples, we will see that the formula that we can derive cannot fit the data perfectly. The aim will be to derive a formula that fits the data best, so that the error between the prediction and the actual data is minimized.

Analysis using R:

We use the statistical analysis software R to calculate the linear dependence relation between the variables degrees Celsius and degrees Fahrenheit.

The R package has the function *lm* which calculates the linear relationship between the variables. It can be used in the following form: *lm(y ~ x, data = dataset_for_x_y)*, where *y* is the variable dependent on *x*. The data frame temperatures should contain the vectors with the values for *x* and *y*:

Input:

```
# source_code/6/frahrenheit_celsius.r
temperatures = data.frame(
    fahrenheit = c(5,14,23,32,41,50), celsius = c(-15,-10,-5,0,5,10)
)
```

```
model = lm(celsius ~ fahrenheit, data = temperatures)
print(model)
```

Output:

```
$ Rscript fahrenheit_celsius.r
Call:
lm(formula = celsius ~ fahrenheit, data = temperatures)
Coefficients: (Intercept)     fahrenheit
                 -17.7778         0.5556
```

Therefore, we can see the following approximate linear dependence relation between *C* (degrees Celsius) and *F* (degrees Fahrenheit):

$$C=fahrenheit*F+Intercept=0.5556*F-17.7778$$

Note that this agrees with our previous calculation.

Visualization:

We display the linear model predicting degrees Celsius from degrees Fahrenheit underneath by a linear line. Its meaning is that the point *(F,C)* is on the green line if and only if *F* (degrees Fahrenheit) converts to *C* (degrees Celsius) and vice versa:

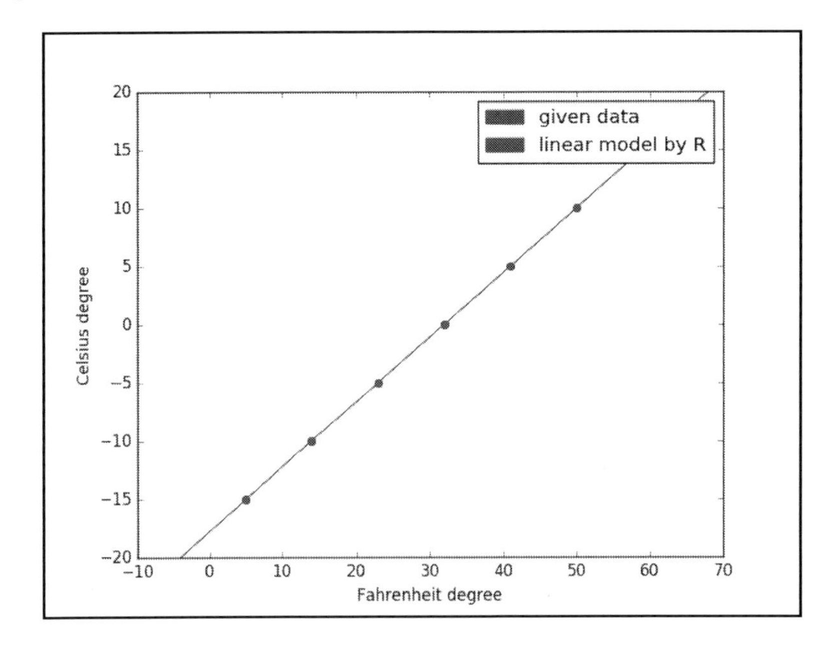

Weight prediction from height - linear regression on real-world data

Here we predict the weight of a man from his height using linear regression from the following data in the table for men:

Height in cm	Weight in kg
180	75
174	71
184	83
168	63
178	70
172	?

We would like to estimate the weight of a man given that his height is 172cm.

Analysis using R:

In the previous example Fahrenheit and Celsius conversion, the data fitted the linear model perfectly. Thus we could perform even a simple mathematical analysis (solving basic equations) to gain the conversion formula. Most of the data in the realworld does not fit a model perfectly. For such an analysis, it is good to find the model that fits the given data with the minimal error. We use R do find such a linear model.

Input:

We put the data from the table above into the vectors and try to fit the linear model.

```
# source_code/6/weight_prediction.r
men = data.frame(
    height = c(180,174,184,168,178), weight = c(75,71,83,63,70)
)
model = lm(weight ~ height, data = men)
print(model)
```

Output:

```
$ Rscript weight_prediction.r
Call:
lm(formula = weight ~ height, data = men)
Coefficients: (Intercept)      height
                 -127.688       1.132
```

Thus the formula expressing the linear relationship between the weight and the height is as follows: *weight=1.132*height-127.688*. Therefore, we estimate that the man with the height of 172cm would have the weight *1.132*172-127.688=67.016 kg*.

Gradient descent algorithm and its implementation

To understand better how we may be able to predict a value using linear regression from first principles, we study a gradient descent algorithm and then implement it in Python.

Gradient descent algorithm

A gradient descent algorithm is an iterative algorithm updating the variables in the model to fit the data with the least error. More generally, it finds a minimum of a function.

We would like to express the weight in terms of the height using a linear formula:

$$weight(height,p)=p_1*height+p_0$$

We estimate the parameter $p=(p_0,p_1)$ using n data samples $(height_i,weight_i)$ to minimize the following square error:

$$E(p) = \frac{1}{2}\sum_{i=1}^{n}[\text{weight}(\text{height}_i, p) - \text{weight}_i]^2$$

The gradient descent algorithm does it by updating the parameter p_i in the direction of $(\partial/\partial p_j) E(p)$, in particular:

$$p_j = p_j - \text{learning\_rate} * \left(\frac{\partial}{\partial p_j} E(p)\right)$$

Here, *learning_rate* determines the speed of the convergence of the *E(p)* to the minimum. Updating of the parameter *p* will result in the convergence of *E(p)* to a certain value providing that *learning_rate* is sufficiently small. In the Python program, we use *learning_rate* of 0.000001. However, the drawback of this update rule is that the minimum of *E(p)* may be only a local minimum.

To update the parameter *p* programatically, we need to unfold the partial derivative on *E(p)*. Therefore, we update the parameter *p* as follows:

$$p_0 := p_0 + \text{learning\_rate} * \sum_{i=1}^{n} [\text{weight}_i - \text{weight}(\text{height}_i, p)]$$

$$p_1 := p_1 + \text{learning\_rate} * \sum_{i=1}^{n} [(\text{weight}_i - \text{weight}(\text{height}_i, p)) * \text{height}_i]$$

We will keep updating the parameter p until it changes only a very little, that is, the change of both p_0 and p_1 is less than some constant *acceptable_error*. Once the parameter *p* stabilizes, we can use it to estimate the weight from the height.

Implementation:

```
# source_code/6/regression.py
# Linear regression program to learn a basic linear model.
import math
import sys
sys.path.append('../common')
import common # noqa

# Calculate the gradient by which the parameter should be updated.
def linear_gradient(data, old_parameter):
    gradient = [0.0, 0.0]
    for (x, y) in data:
        term = float(y) - old_parameter[0] - old_parameter[1] * float(x)
        gradient[0] += term
        gradient[1] += term * float(x)
    return gradient
```

```
# This function will apply gradient descent algorithm
# to learn the linear model.
def learn_linear_parameter(data, learning_rate,
                           acceptable_error, LIMIT):
    parameter = [1.0, 1.0]
    old_parameter = [1.0, 1.0]
    for i in range(0, LIMIT):
        gradient = linear_gradient(data, old_parameter)
        # Update the parameter with the Least Mean Squares rule.
        parameter[0] = old_parameter[0] + learning_rate * gradient[0]
        parameter[1] = old_parameter[1] + learning_rate * gradient[1]
        # Calculate the error between the two parameters to compare with
        # the permissible error in order to determine if the calculation
        # is suffiently accurate.
        if abs(parameter[0] - old_parameter[0]) <= acceptable_error
        and abs(parameter[1] - old_parameter[1]) <= acceptable_error:
            return parameter
        old_parameter[0] = parameter[0]
        old_parameter[1] = parameter[1]
    return parameter

# Calculate the y coordinate based on the linear model predicted.
def predict_unknown(data, linear_parameter):
    for (x, y) in data:
        print(x, linear_parameter[0] + linear_parameter[1] * float(x))

# Program start
csv_file_name = sys.argv[1]
# The maximum number of the iterations in the batch learning algorithm.
LIMIT = 100
# Suitable parameters chosen for the problem given.
learning_rate = 0.0000001
acceptable_error = 0.001

(heading, complete_data, incomplete_data,
 enquired_column) = common.csv_file_to_ordered_data(csv_file_name)
linear_parameter = learn_linear_parameter(
    complete_data, learning_rate, acceptable_error, LIMIT)
print("Linear model:\n(p0,p1)=" + str(linear_parameter) + "\n")
print("Unknowns based on the linear model:")
predict_unknown(incomplete_data, linear_parameter)
```

Input:

We use the data from the table in example weight prediction from height and save it in a CSV file.

```
# source_code/6/height_weight.csv
height,weight
180,75
174,71
184,83
168,63
178,70
172,?
```

Output:

```
$ python regression.py height_weight.csv
Linear model:
(p0,p1)=[0.9966468959362077, 0.4096393414704317]

Unknowns based on the linear model:
('172', 71.45461362885045)
```

The output for the linear model means that the weight can be expressed in terms of the height as follows:

*weight = 0.4096393414704317 * height + 0.9966468959362077*

Therefore, a man with a height of 172cm is predicted to weigh *0.4096393414704317 * 172 + 0.9966468959362077 = 71.45461362885045 ~ 71.455kg.*

Note that this prediction of 71.455kg is slightly different from the prediction in R of 67.016kg. This may be due to the fact that the Python algorithm found only a local minimum in the prediction or that R uses a different algorithm or its implementation.

Visualization - comparison of models by R and gradient descent algorithm

For example, weight prediction from height, we visualize the linear prediction models of R and of the gradient descent algorithm implemented in Python.

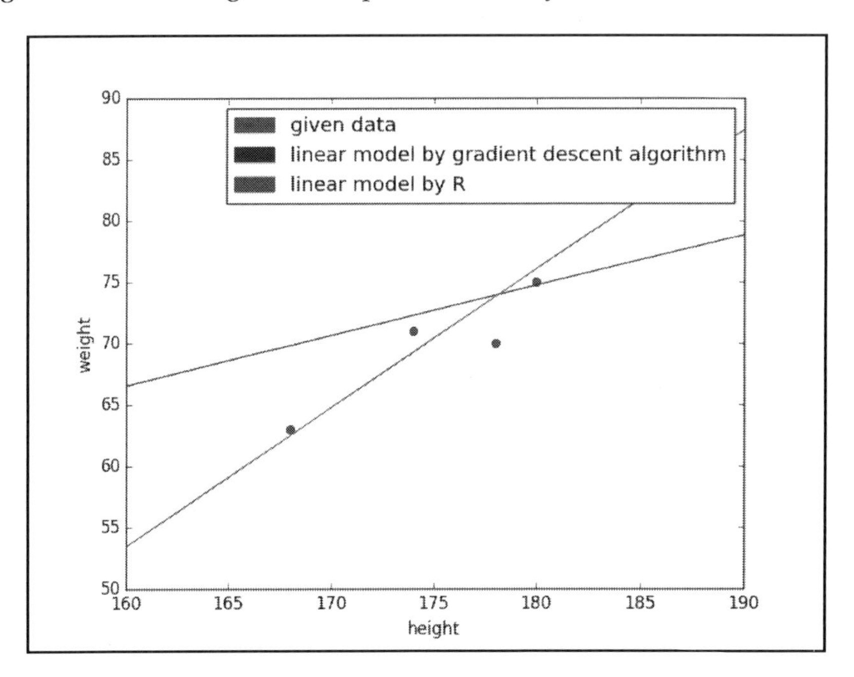

Flight time duration prediction from distance

Given a table of flights with their origin, destination, and flight time, we would like to estimate the length of a proposed flight from Bratislava, Slovakia to Amsterdam, the Netherlands:

Origin	Destination	Distance in km	Flight duration	Flight duration in hours
London	Amsterdam	365	1h 10m	1.167
London	Budapest	1462	2h 20m	2.333
London	Bratislava	1285	2h 15m	2.250
Bratislava	Paris	1096	2h 5m	2.083

Bratislava	Berlin	517	1h 15m	2.250
Vienna	Dublin	1686	2h 50m	2.833
Vienna	Amsterdam	932	1h 55m	1.917
Amsterdam	Budapest	1160	2h 10m	2.167
Bratislava	Amsterdam	978	?	?

Analysis:

We can reason that the flight duration time consists of two times - the first is the time to take off and the landing time; the second is the time that the airplane moves at a certain speed in the air. The first time is some constant. The second time depends linearly on the speed of the plane, which we assume is similar across all the flights in the table. Therefore, the flight time can be expressed using a linear formula in terms of the flight distance.

Analysis using R:

Input:

```
source_code/6/flight_time.r
flights = data.frame(
    distance = c(365,1462,1285,1096,517,1686,932,1160),
    time = c(1.167,2.333,2.250,2.083,2.250,2.833,1.917,2.167)
)
model = lm(time ~ distance, data = flights) print(model)
```

Output:

```
$ Rscript flight_time.r
Call:
lm(formula = time ~ distance, data = flights)
Coefficients: (Intercept)     distance
                1.2335890    0.0008387
```

According to the linear regression, the time to take off and the landing time for an average flight is about 1.2335890 hours. Then to travel 1 km with the plane takes 0.0008387 hours; in other words, the speed of an airplane is 1192 km per hour. The actual usual speed of an aeroplane for short-distance flights like the ones in the table is about 850 km per hour. This leaves room for improvement in our estimation (refer to exercise 6.3).

Therefore, we derived the following formula:

*flight_time=0.0008387*distance+1.2335890 hours*

Using it, we estimate that the flight from Bratislava to Amsterdam, with the distance 978 km, would take about *0.0008387*978+1.2335890=2.0538376* hours or about 2 hours and 3 minutes, which is a little longer than from Vienna to Amsterdam (1h 55m) and a little shorter than from Budapest to Amsterdam (2h 10m).

Ballistic flight analysis – non-linear model

An interplanetary spaceship lands on a planet with negligible atmosphere and fires three projectiles at the same angle carrying exploratory bots, but at different initial velocities. After the bots land on the surface their distances are measured and the data recorded as follows:

Velocity in m/s	Distance in m
400	38 098
600	85 692
800	152 220
?	300 000

At what speed should the projectile carrying the 4th bot be fired in order for it to land 300 km from the spacecraft?

Analysis:

For this problem we need to understand the trajectory of the projectile. Since the atmosphere on the explored planet is weak, the trajectory is almost equivalent to the ballistic curve without the air drag. The distance d traveled by an object fired from a point on the ground is approximately (neglecting the curving of the planet surface) given by the equation:

$$d = v^2 * \sin(2 * \tau)/g$$

Where v is the initial velocity of the object, τ is an angle at which the object was fired and g is the gravitational force exerted by the planet on the object. Note that the angle τ and the gravitational force g do not change. Therefore define a constant $c = \sin(2 * \tau)/g$. Then the distance on the explored planet can be explained in terms of the velocity by the equation:

$$d = v^2 * c$$

Although d and v are not in the linear relationship, d and the square of v are. Therefore we can still apply the linear regression to determine the relationship between d and v.

Analysis using R:

Input:

```
source_code/6/speed_distance.r
trajectories = data.frame(
    squared_speed = c(160000,360000,640000),
    distance = c(38098, 85692, 152220)
)
model = lm(squared_speed ~ distance, data = trajectories)
print(model)
```

Output:

```
$ Rscript speed_distance.r
Call:
lm(formula = squared_speed ~ distance, data = trajectories)
Coefficients:
(Intercept)     distance
   -317.708        4.206
```

Therefore the relationship between the squared velocity and the distance is predicted by the regression to be:

$$v^2 = 4.206 * d - 317.708.$$

The presence of the intercept term may be caused by the errors in the measurements or by other forces playing in the equation. Since it is relatively small, the final velocity should be estimated reasonably well. Putting the distance of 300km into the equation we get:

$$v^2 = 4.206 * 300000 - 317.708=1261482.292$$

$$v=1123.157$$

Therefore for the projectile to reach the 300km from the source, we need to fire it at the speed of 1123.157 m/s approximately.

Summary

We can think of variables as being dependent on each other in a functional way. For example, the variable y is a function of x denoted by $y=f(x)$. The function $f(x)$ has constant parameters. For example, if y depends on x linearly, then $f(x)=a*x+b$, where a and b are constant parameters in the function $f(x)$. Regression is a method to estimate these constant parameters in such a way that the estimated $f(x)$ follows y as closely as possible. This is formally measured by the squared error between $f(x)$ and y for the data samples x.

The gradient descent method minimizes this error by updating the constant parameters in the direction of the steepest descent (that is, the partial derivative of the error), ensuring that the parameters converge to the values resulting in the minimal error in the quickest possible way.

The statistical software R supports the estimation of the linear regression with the function *lm*.

Problems

1. **Cloud storage prediction cost**: Our software application generates data on a monthly basis and stores this data in cloud storage together with the data from the previous months. We are given the following bills for the cloud storage and we would like to estimate the running costs for the first year of using this cloud storage:

Month of using the cloud storage	Monthly bill in euros
1	120.0
2	131.2
3	142.1
4	152.9
5	164.3
1 to 12	?

2. **Fahrenheit and Celsius conversion**: In the earlier example, we devised a formula converting degrees Fahrenheit into degrees Celsius. Devise a formula converting degrees Celsius into degrees Fahrenheit.
3. **Flight time duration prediction from the distance**: Why do you think that a linear regression model resulted in the estimation of the speed to be 1192 km/h as opposed to the real speed of about 850 km/h? Can you suggest a way to a better model of the estimation of the flight duration based on the flight distances and times?
4. **Bacteria population prediction**: A bacteria Escherichia coli has been observed in the laboratory and the size of its population was estimated by various measurements at 5-minute intervals as follows:

Time	Size of population in millions
10:00	47.5
10:05	56.5
10:10	67.2
10:15	79.9
11:00	?

What is the expected number of the bacteria to be observed at 11:00 assuming that the bacteria would continue to grow at the same rate?

Analysis:

1. Every month, we have to pay for the data we have stored in the cloud storage so far plus for the new data that is added to the storage in that month. We will use linear regression to predict the cost for a general month and then we will calculate the sum of the first 12 months to calculate the cost for the whole year.

Input:

```
source_code/6/cloud_storage.r
bills = data.frame(
    month = c(1,2,3,4,5),
    bill = c(120.0,131.2,142.1,152.9,164.3)
)
model = lm(bill ~ month, data = bills) print(model)
```

Output:

```
$ Rscript cloud_storage.r
Call:
lm(formula = bill ~ month, data = bills)
Coefficients: (Intercept)      month
                    109.01     11.03
```

This means that the base cost is *base_cost=109.01* euros and then to store the data added in 1 month costs additional *month_data=11.03* euros. Therefore the formula for the nth monthly bill is as follows:

*bill_amount=month_data*month_number+base_cost=11.03*month_number+109.01 euro*

Remember that the sum of the first n numbers is *(1/2)*n*(n+1)*. Thus the cost for the first *n* months will be as follows:

*total_cost(n months)=base_cost*n+month_data*[(1/2)*n*(n+1)]*

=n[base_cost+month_data*(1/2)*(n+1)]*

=n[109.01+11.03*(1/2)*(n+1)]*

=n[114.565+5.515*n]*

Thus for the whole year, the cost will be as follows:

total_cost(12 months)=12[114.565+5.515*12]=2168.94 euros*

Visualization:

In the graph below, we can observe the linearity of the model represented by the blue line. On the other hand, the sum of the points on the linear line is quadratic in nature and is represented by the area under the line.

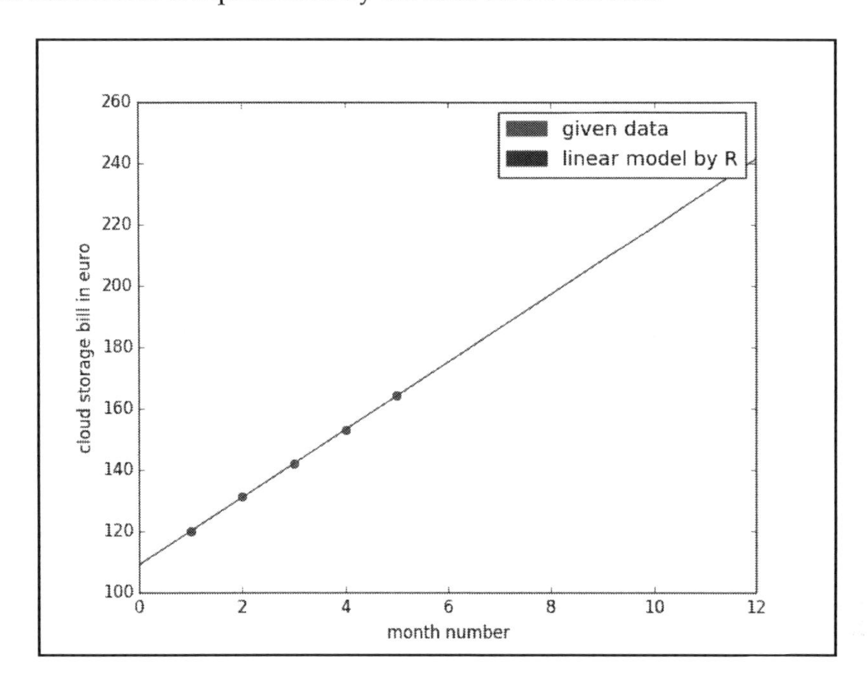

2. There are many ways to obtain the formula converting degrees Celsius into degrees Fahrenheit. We could use R and from the initial R file take the following line:

   model = lm(celsius ~ fahrenheit, data = temperatures)

   We then change it to:

   model = lm(fahrenheit ~ celsius, data = temperatures)

Then we would obtain the desired reversed model:

```
Call:
lm(formula = fahrenheit ~ celsius, data = temperatures)
Coefficients:
(Intercept)    celsius
       32.0        1.8
```

So degrees Fahrenheit can be expressed from degrees Celsius as: $F=1.8*C+32$.

We may obtain this formula alternatively by modifying the formula:

$C=(5/9)*F-160/9$

$160/9+C=(5/9)*F$

$160+9*C=5*F$ $F=1.8*C+32$

3. The estimated speed is so high because even flights over a short distance take quite long: for example, the flight from London to Amsterdam, where the distance between the two cities is only 365 km, takes about 1.167 hours. But, on the other hand, if the distance changes only a little, then the flight time changes only a little as well. This results in us estimating a very high initial setup time. Consequently, the speed has to be very high because there is only a small amount of time left to travel a certain distance.

    If we consider very long flights where the initial setup time to flight time ratio is much smaller, we could predict the flight speed more accurately.

4. The number of the bacteria at the 5-minute intervals is: 47.5, 56.5, 67.2, and 79.9 millions. The differences between these numbers are: 9, 10.7, and 12.7. The sequence is increasing. So we look at the ratios of the neighbor terms to see how the sequence grows. 56.5/47.5=1.18947, 67.2/56.5=1.18938, and 79.9/67.2=1.18899. The ratios of the successive terms are close to each other, so we have the reason to believe that the number of the bacteria in the growing population can be estimated using the exponential distribution by the model:

$$n = 47.7 * b^m$$

Where n is the number of the bacteria in millions, b is a constant (the base), the number m is the exponent expressing the number of the minutes since 10:00 which is the time of the first measurement, 47.7 is the number of the bacteria at this measurement in millions.

To estimate the constant b, we use the ratios between the sequence terms. We know that b^5 is approximately (56.5/47.5 + 67.2/56.5 + 79.9/67.2)/3=1.18928. Therefore the constant b is approximately $b=1.18928^{1/5}=1.03528$. Thus the number of the bacteria in millions is:

$$n = 47.7 * 1.03528^m$$

At 11:00, which is 60 minutes later than 10:00, the estimated number of bacteria is:

$47.7*1.03528^{60}=381.9$ $7.7*1.03528^{60}=381.9$ million.

7
Time Series Analysis

Time series analysis is the analysis of time-dependent data. Given data for a certain period, the aim is to predict data for a different period, usually in the future. For example, time series analysis is used to predict financial markets, earthquakes, and weather. In this chapter, we are mostly concerned with predicting the numerical values of certain quantities, for example, the human population in 2030.

The main elements of time-based prediction are:

- The trend of the data: does the variable tend to rise or fall as time passes? For example, does human population grow or shrink?
- Seasonality: how is the data dependent on certain regular events in time? For example, are restaurant sales bigger on Fridays than on Tuesdays?

Combining these two elements of time series analysis equips us with a powerful method to make time-dependent predictions. In this chapter, you will learn the following:

- How to analyse data trends using regression in an example business profits
- How to observe and analyse recurring patterns in data in a form of seasonality in an example about an Electronics shop's sales
- Using the example of an electronics shop's sales, to combine the analysis of trends and seasonality to predict time-dependent data
- Create time-dependent models in R using the examples of business profits and an electronics shop's sales

Business profit - analysis of the trend

We are interested in predicting the profits of a business for the year 2018 given its profits for the previous years:

Year	Profit in USD
2011	40k
2012	43k
2013	45k
2014	50k
2015	54k
2016	57k
2017	59k
2018	?

Analysis:

In this example, the profit is always increasing, so we can think of representing the profit as a growing function dependent on the time variable represented by years. The differences in profit between the subsequent years are: 3k, 2k, 5k, 4k, 3k, and 2k USD. These differences do not seem to be affected by time, and the variation between them is relatively low. Therefore, we may try to predict the profit for the coming years by performing a linear regression. We express profit p in terms of the year y in the linear equation, also called a trend line:

*profit=a*year+b*

We can find the constants *a* and *b* with linear regression.

Input:

We store the data from the table above in the vectors year and profit in R script.

```
# source_code/7/profit_year.r
business_profits = data.frame(
    year = c(2011,2012,2013,2014,2015,2016,2017),
    profit = c(40,43,45,50,54,57,59)
)
model = lm(profit ~ year, data = business_profits)
print(model)
```

Output:

```
$ Rscript profit_year.r
Call:
lm(formula = profit ~ year, data = business_profits)
Coefficients:
(Intercept)    year
  -6711.571   3.357
```

Visualization:

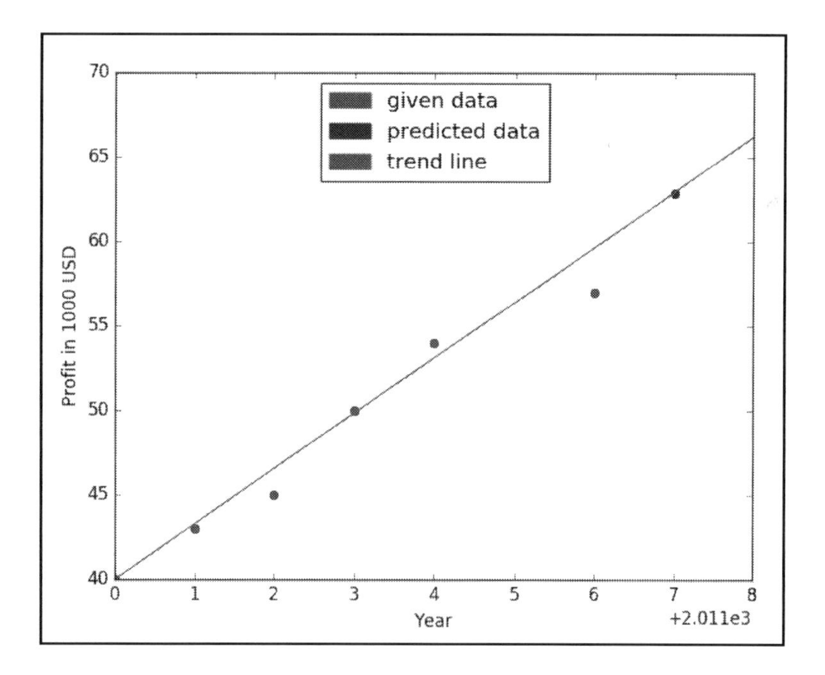

Conclusion:

Therefore, the trend line equation for the profit of the company is:
*profit=3.357*year-6711.571*.

From this equation, we can predict the profit for the year 2018 to be
*profit=3.357*2018-6711.571=62.855k USD or 62855 USD*.

This example was simple - we were able to make a prediction just by using linear regression on the trend line. In the next example, we will look at data subject to both trends and seasonality.

Electronics shop's sales - analysis of seasonality

We have data of sales in thousands of USD for a small electronics shop by month for the years 2010 to 2017. We would like to predict sales for each month of 2018:

Month/Year	2010	2011	2012	2013	2014	2015	2016	2017	2018
January	10.5	11.9	13.2	14.6	15.1	16.5	18.9	20	20.843
February	11.9	12.6	14.4	15.4	17.4	17.9	19.5	20.8	21.993
March	13.4	13.5	16.1	16.2	17.2	19.6	19.8	22.1	22.993
April	12.7	13.6	14.9	17.8	17.8	20.2	19.7	20.9	22.956
May	13.9	14.6	15.7	17.8	18.6	19.1	20.8	21.5	23.505
June	14	14.4	15.3	16.1	18.9	19.7	21.1	22.1	23.456
July	13.5	15.7	16.8	17.4	18.3	19.7	21	22.6	23.881
August	14.5	14	15.7	17	17.9	20.5	21	22.7	23.668
September	14.3	15.5	16.8	17.2	19.2	20.3	20.6	21.9	23.981
October	14.9	15.8	16.3	17.9	18.8	20.3	21.4	22.9	24.293
November	16.9	16.5	18.7	20.5	20.4	22.4	23.7	24	26.143
December	17.4	20.1	19.7	22.5	23	23.8	24.6	26.6	27.968

Analysis:

To be able to analyze this, we will first graph the data so that we can notice patterns and analyze them.

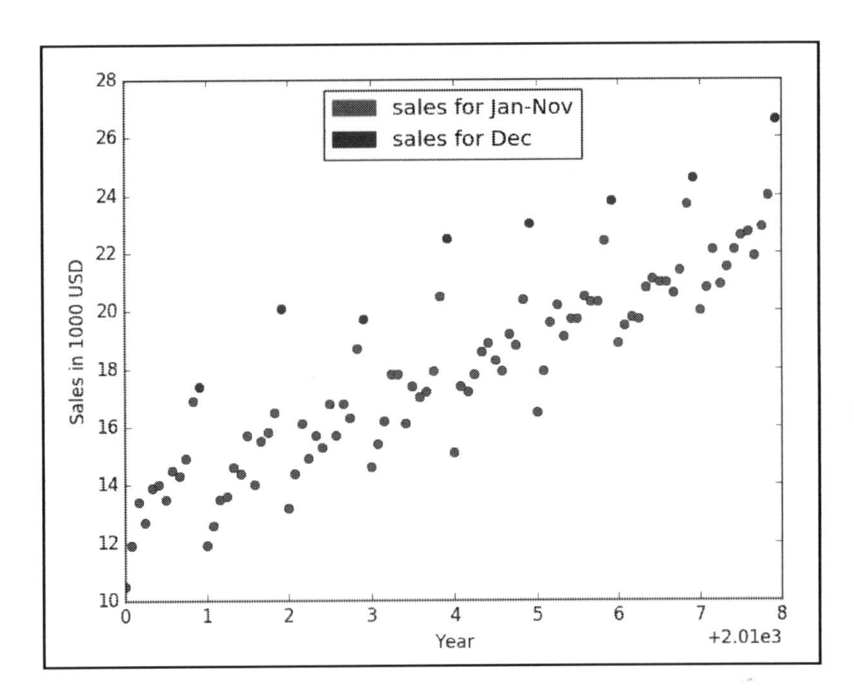

From the graph and the table, we notice that, in the long term, the sales increase linearly. This is the trend. However, we can also see that the sales for December tend to be higher than for the other months. Thus, we have reason to believe that sales are also influenced by the month.

How could we predict the monthly sales for the following years? First, we determine the exact long-term trend of the data. Then, we would like to analyze the change across the months.

Analyzing trends using R

Input:

The year list contains the periods of the year represented as a decimal number *year+month/12*. The sales list contains the sales in thousands of USD for the corresponding periods in the year list. We will use linear regression to find the trend line. From the initial graph, we notice that the trend is linear in nature.

```
# source_code/6/sales_year.r
#Predicting sales based on the period in the year
sales = data.frame(
    year = c(2010.000000, 2010.083333, 2010.166667, 2010.250000,
             2010.333333, 2010.416667, 2010.500000, 2010.583333,
             2010.666667, 2010.750000, 2010.833333, 2010.916667,
             2011.000000, 2011.083333, 2011.166667, 2011.250000,
             2011.333333, 2011.416667, 2011.500000, 2011.583333,
             2011.666667, 2011.750000, 2011.833333, 2011.916667,
             2012.000000, 2012.083333, 2012.166667, 2012.250000,
             2012.333333, 2012.416667, 2012.500000, 2012.583333,
             2012.666667, 2012.750000, 2012.833333, 2012.916667,
             2013.000000, 2013.083333, 2013.166667, 2013.250000,
             2013.333333, 2013.416667, 2013.500000, 2013.583333,
             2013.666667, 2013.750000, 2013.833333, 2013.916667,
             2014.000000, 2014.083333, 2014.166667, 2014.250000,
             2014.333333, 2014.416667, 2014.500000, 2014.583333,
             2014.666667, 2014.750000, 2014.833333, 2014.916667,
             2015.000000, 2015.083333, 2015.166667, 2015.250000,
             2015.333333, 2015.416667, 2015.500000, 2015.583333,
             2015.666667, 2015.750000, 2015.833333, 2015.916667,
             2016.000000, 2016.083333, 2016.166667, 2016.250000,
             2016.333333, 2016.416667, 2016.500000, 2016.583333,
             2016.666667, 2016.750000, 2016.833333, 2016.916667,
             2017.000000, 2017.083333, 2017.166667, 2017.250000,
             2017.333333, 2017.416667, 2017.500000, 2017.583333,
             2017.666667, 2017.750000, 2017.833333, 2017.916667),
    sale = c(10.500000, 11.900000, 13.400000, 12.700000, 13.900000,
             14.000000, 13.500000, 14.500000, 14.300000, 14.900000,
             16.900000, 17.400000, 11.900000, 12.600000, 13.500000,
             13.600000, 14.600000, 14.400000, 15.700000, 14.000000,
             15.500000, 15.800000, 16.500000, 20.100000, 13.200000,
             14.400000, 16.100000, 14.900000, 15.700000, 15.300000,
             16.800000, 15.700000, 16.800000, 16.300000, 18.700000,
             19.700000, 14.600000, 15.400000, 16.200000, 17.800000,
             17.800000, 16.100000, 17.400000, 17.000000, 17.200000,
             17.900000, 20.500000, 22.500000, 15.100000, 17.400000,
             17.200000, 17.800000, 18.600000, 18.900000, 18.300000,
```

```
                17.900000, 19.200000, 18.800000, 20.400000, 23.000000,
                16.500000, 17.900000, 19.600000, 20.200000, 19.100000,
                19.700000, 19.700000, 20.500000, 20.300000, 20.300000,
                22.400000, 23.800000, 18.900000, 19.500000, 19.800000,
                19.700000, 20.800000, 21.100000, 21.000000, 21.000000,
                20.600000, 21.400000, 23.700000, 24.600000, 20.000000,
                20.800000, 22.100000, 20.900000, 21.500000, 22.100000,
                22.600000, 22.700000, 21.900000, 22.900000, 24.000000,
                26.600000)
)

model = lm(sale ~ year, data = sales)
print(model)
```

Output:

```
$ Rscript sales_year.r
Call:
lm(formula = sale ~ year, data = sales)
Coefficients: (Intercept)      year
               -2557.778     1.279
```

Therefore, the equation of the trend line is:

*sales = 1.279*year-2557.778*

Visualization:

Now we add the trend line to the graph:

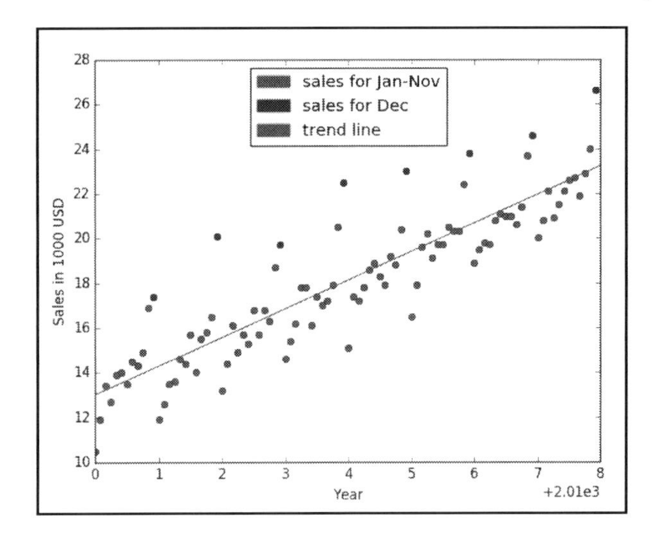

Analyzing seasonality

Now we analyze seasonality - how data changes across months. From our observations, we know that, for some months, sales tend to be higher, whereas, for other months, sales tend to be lower. We evaluate the differences between the linear trend and the actual sales. Based on the pattern observed in these differences, we produce a model of seasonality to predict sales more accurately for each month:

Sales for January									
Year	2010	2011	2012	2013	2014	2015	2016	2017	Average
Actual sales	10.5	11.9	13.2	14.6	15.1	16.5	18.9	20	
Sales on the trend line	13.012	14.291	15.57	16.849	18.128	19.407	20.686	21.965	
Difference	-2.512	-2.391	-2.37	-2.249	-3.028	-2.907	-1.786	-1.965	-2.401
Sales for February									
Year	2010	2011	2012	2013	2014	2015	2016	2017	Average
Actual sales	11.9	12.6	14.4	15.4	17.4	17.9	19.5	20.8	
Sales on the trend line	13.1185833333	14.3975833333	15.6765833333	16.9555833333	18.2345833333	19.5135833333	20.7925833333	22.0715833333	
Difference	-1.2185833333	-1.7975833333	-1.2765833333	-1.5555833333	-0.8345833333	-1.6135833333	-1.2925833333	-1.2715833333	-1.3575833333
Sales for March									
Year	2010	2011	2012	2013	2014	2015	2016	2017	Average
Actual sales	13.4	13.5	16.1	16.2	17.2	19.6	19.8	22.1	
Sales on the trend line	13.2251666667	14.5041666667	15.7831666667	17.0621666667	18.3411666667	19.6201666667	20.8991666667	22.1781666667	
Difference	0.1748333333	-1.0041666667	0.3168333333	-0.8621666667	-1.1411666667	-0.0201666667	-1.0991666667	-0.0781666667	-0.4641666667
Sales for April									
Year	2010	2011	2012	2013	2014	2015	2016	2017	Average
Actual sales	12.7	13.6	14.9	17.8	17.8	20.2	19.7	20.9	
Sales on the trend line	13.33175	14.61075	15.88975	17.16875	18.44775	19.72675	21.00575	22.28475	
Difference	-0.63175	-1.01075	-0.98975	0.63125	-0.64775	0.47325	-1.30575	-1.38475	-0.60825
Sales for May									
Year	2010	2011	2012	2013	2014	2015	2016	2017	Average
Actual sales	13.9	14.6	15.7	17.8	18.6	19.1	20.8	21.5	

Sales on the trend line	13.4383333333	14.7173333333	15.9963333333	17.2753333333	18.5543333333	19.8333333333	21.1123333333	22.3913333333	
Difference	0.4616666667	-0.1173333333	-0.2963333333	0.5246666667	0.0456666667	-0.7333333333	-0.3123333333	-0.8913333333	-0.1648333333
Sales for June									
Year	2010	2011	2012	2013	2014	2015	2016	2017	Average
Actual sales	14	14.4	15.3	16.1	18.9	19.7	21.1	22.1	
Sales on the trend line	13.5449166667	14.8239166667	16.1029166667	17.3819166667	18.6609166667	19.9399166667	21.2189166667	22.4979166667	
Difference	0.4550833333	-0.4239166667	-0.8029166667	-1.2819166667	0.2390833333	-0.2399166667	-0.1189166667	-0.3979166667	-0.3214166667
Sales for July									
Year	2010	2011	2012	2013	2014	2015	2016	2017	Average
Actual sales	13.5	15.7	16.8	17.4	18.3	19.7	21	22.6	
Sales on the trend line	13.6515	14.9305	16.2095	17.4885	18.7675	20.0465	21.3255	22.6045	
Difference	-0.1515	0.7695	0.5905	-0.0885	-0.4675	-0.3465	-0.3255	-0.0045	-0.003
Sales for August									
Year	2010	2011	2012	2013	2014	2015	2016	2017	Average
Actual sales	14.5	14	15.7	17	17.9	20.5	21	22.7	
Sales on the trend line	13.7580833333	15.0370833333	16.3160833333	17.5950833333	18.8740833333	20.1530833333	21.4320833333	22.7110833333	
Difference	0.7419166667	-1.0370833333	-0.6160833333	-0.5950833333	-0.9740833333	0.3469166667	-0.4320833333	-0.0110833333	-0.3220833333
Sales for September									
Year	2010	2011	2012	2013	2014	2015	2016	2017	Average
Actual sales	14.3	15.5	16.8	17.2	19.2	20.3	20.6	21.9	
Sales on the trend line	13.8646666667	15.1436666667	16.4226666667	17.7016666667	18.9806666667	20.2596666667	21.5386666667	22.8176666667	
Difference	0.4353333333	0.3563333333	0.3773333333	-0.5016666667	0.2193333333	0.0403333333	-0.9386666667	-0.9176666667	-0.1161666667
Sales for October									
Year	2010	2011	2012	2013	2014	2015	2016	2017	Average
Actual sales	14.9	15.8	16.3	17.9	18.8	20.3	21.4	22.9	
Sales on the trend line	13.97125	15.25025	16.52925	17.80825	19.08725	20.36625	21.64525	22.92425	
Difference	0.92875	0.54975	-0.22925	0.09175	-0.28725	-0.06625	-0.24525	-0.02425	0.08975
Sales for November									
Year	2010	2011	2012	2013	2014	2015	2016	2017	Average

Actual sales	16.9	16.5	18.7	20.5	20.4	22.4	23.7	24	
Sales on the trend line	14.0778333333	15.3568333333	16.6358333333	17.9148333333	19.1938333333	20.4728333333	21.7518333333	23.0308333333	
Difference	2.8221666667	1.1431666667	2.0641666667	2.5851666667	1.2061666667	1.9271666667	1.9481666667	0.9691666667	1.8331666667
Sales for December									
Year	2010	2011	2012	2013	2014	2015	2016	2017	Average
Actual sales	17.4	20.1	19.7	22.5	23	23.8	24.6	26.6	
Sales on the trend line	14.1844166667	15.4634166667	16.7424166667	18.0214166667	19.3004166667	20.5794166667	21.8584166667	23.1374166667	
Difference	3.2155833333	4.6365833333	2.9575833333	4.4785833333	3.6995833333	3.2205833333	2.7415833333	3.4625833333	3.5515833333

We cannot observe any obvious trends in the differences between actual sales and sales on the trend line. Therefore, we just calculate the arithmetic means of these differences for every month.

For example, we notice that sales in December tend to be higher by about 3551.58 USD compared to sales predicted on the trend line. Similarly, sales for January tend to be lower on average by 2401 USD compared to sales predicted on the trend line.

Making the assumption that the month has an impact on the actual sales from our observations of the variation of sales across the months, we take our prediction rule:

$$sales = 1.279*year - 2557.778$$

We then update it to the new rule:

$$sales = 1.279*year - 2557.778 + month\_difference$$

Here, sales is the amount of sales for a chosen month and year in the prediction, and *month_difference* is the average difference in our given data between actual sales and sales on the trend line. More specifically, we get the following 12 equations and predictions for sales for the year 2018 in thousands of USD:

$$sales\_january = 1.279*(year+0/12) - 2557.778 - 2.401$$

$$= 1.279*(2018 + 0/12) - 2557.778 - 2.401 = 20.843$$

$$sales\_february = 1.279*(year+1/12) - 2557.778 - 1.358$$

$$= 1.279*(2018+1/12) - 2557.778 - 1.358 = 21.993$$

$$sales\_march = 1.279*(year+2/12) - 2557.778 - 0.464$$

$$= 1.279*(2018+2/12) - 2557.778 - 0.464 = 22.993$$

sales_april = 1.279(year+3/12) - 2557.778 - 0.608*

= 1.279(2018+3/12) - 2557.778 - 0.608 = 22.956*

sales_may = 1.279(year+4/12) - 2557.778 - 0.165*

= 1.279(2018+4/12) - 2557.778 - 0.165 = 23.505*

sales_june = 1.279(year+5/12) - 2557.778 - 0.321*

= 1.279(2018+5/12) - 2557.778 - 0.321 = 23.456*

sales_july = 1.279(year+6/12) - 2557.778 - 0.003*

= 1.279(2018+6/12) - 2557.778 - 0.003 = 23.881*

sales_august = 1.279(year+7/12) - 2557.778 - 0.322*

= 1.279(2018+7/12) - 2557.778 - 0.322 = 23.668*

sales_september = 1.279(year+8/12) - 2557.778 - 0.116*

= 1.279(2018+8/12) - 2557.778 - 0.116 = 23.981*

sales_october = 1.279(year+9/12) - 2557.778 + 0.090*

= 1.279(2018+9/12) - 2557.778 + 0.090 = 24.293*

sales_november = 1.279(year+10/12) - 2557.778 + 1.833*

= 1.279(2018+10/12) - 2557.778 + 1.833 = 26.143*

sales_december = 1.279(year+11/12) - 2557.778 + 3.552*

= 1.279(2018+11/12) - 2557.778 + 3.552 = 27.968*

Conclusion

Therefore, we complete the table with sales for the year 2018 based on the seasonal equations above.

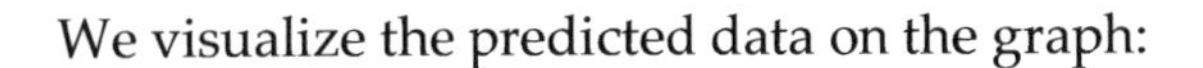

We visualize the predicted data on the graph:

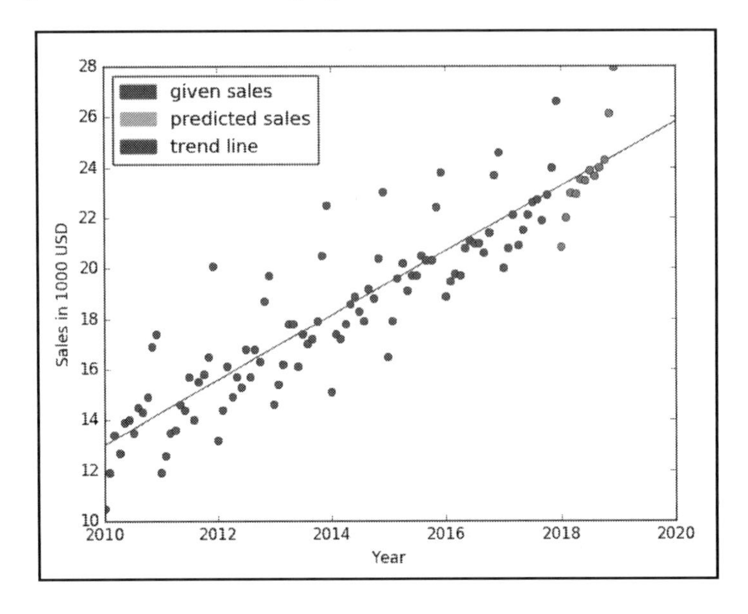

Summary

Time series analysis is the analysis of time-dependent data. The two most important factors in this analysis are the analysis of trends and the analysis of seasonality.

The analysis of trends can be considered as determining the function around which the data is distributed. Using the fact that data is dependent on time, this function can be determined using regression. Many phenomena have a linear trend line, whereas others may not follow a linear pattern.

The analysis of the seasonality tries to detect regular patterns occurring in time repeatedly, such as higher sales before Christmas and so on. To detect a seasonal pattern, it is essential to divide data into the different seasons in such a way that a pattern reoccurs in the same season. This division can divide a year into months, a week into days or into workdays and the weekend, and so on. An appropriate division into seasons and analyzing patterns in those is the key to good seasonal analysis.

Once trend and seasonality have been analyzed in the data, the combined result is a predictor for the pattern that the time-dependent data will follow in the future.

Problems

Determining the trend for Bitcoin prices.

a) We are given the table for the Bitcoin prices for the years 2010 - 2017 in terms of USD. Determine a linear trend line for these prices. The monthly price is for the first day in the month:

Date year-month-day	Bitcoin price in USD
2010-12-01	0.23
2011-06-01	9.57
2011-12-01	3.06
2012-06-01	5.27
2012-12-01	12.56
2013-06-01	129.3
2013-12-01	946.92
2014-06-01	629.02
2014-12-01	378.64
2015-06-01	223.31
2015-12-01	362.73
2016-06-01	536.42
2016-12-01	753.25
2017-06-01	2452.18

Data taken from CoinDesk price page.

b) As per the linear trend line from part **a)**, what is the expected price of Bitcoin in 2020?

c) Discuss whether a linear line is a good indicator for the future price of Bitcoin.

Electronics shop's sales. Using the data in the electronics shop's sales example, predict the sales for every month of the year 2019.

Analysis:

Input:

```
source_code/7/year_bitcoin.r
#Determining a linear trend line for Bitcoin
bitcoin_prices = data.frame(
    year = c(2010.91666666666, 2011.41666666666, 2011.91666666666,
2012.41666666666, 2012.91666666666, 2013.41666666666,
2013.91666666666, 2014.41666666666, 2014.91666666666,
2015.41666666666, 2015.91666666666, 2016.41666666666,
2016.91666666666, 2017.41666666666),
    btc_price = c(0.23, 9.57, 3.06, 5.27, 12.56, 129.3, 946.92, 629.02,
378.64, 223.31, 362.73, 536.42, 753.25, 2452.18)
)
model = lm(btc_price ~ year, data = bitcoin_prices)
print(model)
```

Output:

```
$ Rscript year_bitcoin.r
Call:
lm(formula = btc_price ~ year, data = bitcoin_prices)
Coefficients: (Intercept)     year
               -431962.9    214.7
```

Trend line:

From the output of the Rscript, we find out that the linear trend line for the price of Bitcoin in USD is:

$price = year * 214.7 - 431962.9$

This gives us the following graph for the trend line:

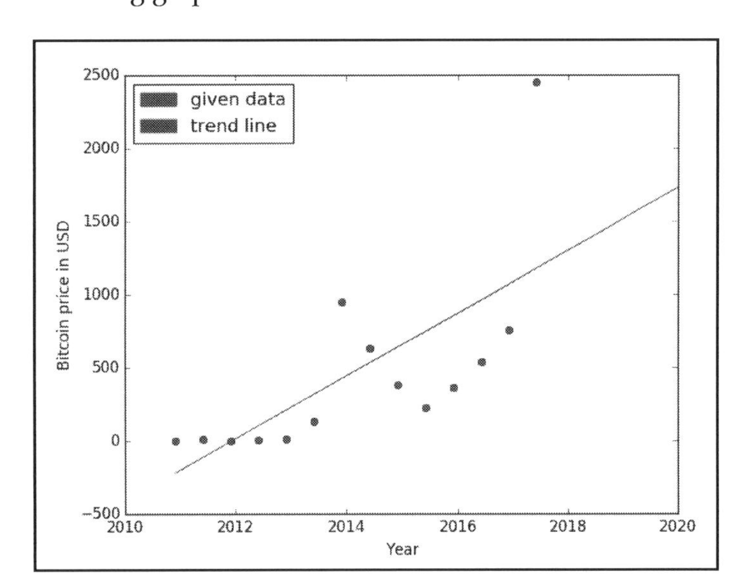

As per the trend line, the expected price for Bitcoin for January 1, 2020 is 1731.1 USD.

A linear trend line is probably not a good indicator and price predictor for Bitcoin. This is because of the many factors in play and because of the potential exponential nature often seen in the trends in technology, for example, the number of active Facebook users and the number of transistors in the best consumer CPU under 1000 USD.

There are three important factors that could facilitate an exponential adoption of Bitcoin and thus drive its price upwards:

- **Technological maturity (scalability)** - the number of transactions per second can ensure an instant transfer, even though many people use Bitcoin to make and receive payments
- **Stability** - once sellers are not afraid to lose their profits if they receive payments in Bitcoin, they are more open to accept it as currency
- **User-friendliness** - once ordinary users can make and receive payments in Bitcoin in a natural way, there will not be a technical barrier to using Bitcoin as they would any other currency they are used to.

To analyze the price of Bitcoin, we would have to take much more data into consideration and it is likely that its price will not follow a linear trend.

We use the 12 formulas from the example, one for each month, to predict the sales for each month in the year 2019:

sales_january = 1.279(year+0/12) - 2557.778 - 2.401*

= 1.279(2019 + 0/12) - 2557.778 - 2.401 = 22.122*

sales_february = 1.279(2019+1/12) - 2557.778 - 1.358 = 23.272*

sales_march = 1.279(2019+2/12) - 2557.778 - 0.464 = 24.272*

sales_april = 1.279(2019+3/12) - 2557.778 - 0.608 = 24.234*

sales_may = 1.279(2019+4/12) - 2557.778 - 0.165 = 24.784*

sales_june = 1.279(2019+5/12) - 2557.778 - 0.321 = 24.735*

sales_july = 1.279(2019+6/12) - 2557.778 - 0.003 = 25.160*

sales_august = 1.279(2019+7/12) - 2557.778 - 0.322 = 24.947*

sales_september = 1.279(2019+8/12) - 2557.778 - 0.116 = 25.259*

sales_october = 1.279(2019+9/12) - 2557.778 + 0.090 = 25.572*

sales_november = 1.279(2019+10/12) - 2557.778 + 1.833 = 27.422*

sales_december = 1.279(2019+11/12) - 2557.778 + 3.552 = 29.247*

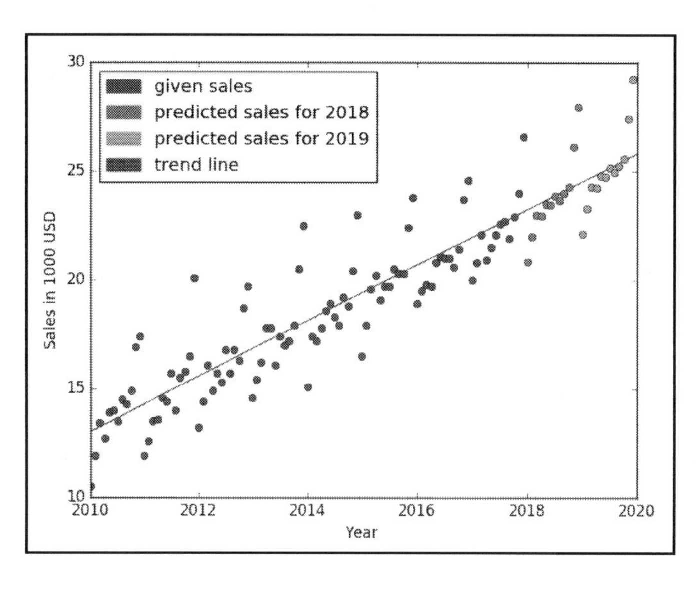

Basic concepts

Notation:

Set intersection of two sets A and B denoted by $A \cap B$ is the subset of A or B that contains all elements that are in both A and B, i.e. $A \cap B$:= { x : x ◎ A and x ◎ B}.

Set union of two sets A and B denoted by $A \cup B$ is the set that contains precisely the elements that are in A or in B, i.e $A \cup B$:= { x : x ◎ A or x ◎ B}.

Set difference of the two sets A and B denoted by $A - B$ or $A \backslash B$ is the subset of A that contains all elements in A that are not in B, i.e. $A - B$:= { x : x ◎ A and x ◎ B}.

Summation symbol $\sum$ represents the sum of all members over the set, e.g.:

$$\sum_{i=1}^{n} a_i = a_1 + a_2 + \ldots + a_n$$

Definitions and terms:

- Population: A set of the similar data or items subject to the analysis.
- Sample: A subset of the population.
- Arithmetic mean (average) of a set: The sum of all the values in the set divided by the size of the set
- Median: The middle value in an ordered set, for example, the median of the set $\{x_1, \ldots, x_{2k+1}\}$ where $x_1 < \ldots < x_{2k+1}$ is the value x_{k+1}.

- Random variable: A function from the set of possible outcomes to the set of the values (for example, real numbers).
- Expectation: An expectation of a random variable is the limit of the average values of the increasing sets of the values given by the random variable.
- Variance: Measures the dispersion of the population from its mean. Mathematically, the variance of a random variable X is the expected value of the square of the difference between the random variable and the mean μ of X, i.e. $Var(X) = E[(X - \mu)^2]$.
- Standard deviation: The deviation of the random variable X is the square root of the variation of the variable X, i.e. *SD(X)=sqrt(Var(X)).*
- Correlation: The measure of the dependency between the random variables. Mathematically, for the random variables X and Y, the correlation is defined as $corr(X,Y) = E[(X - \mu_X) * (Y-\mu_Y)]/(SD(X) * SD(Y))$.
- Causation: A dependence relation explaining the occurrence of one phenomena through the occurrence of another phenomena. Causation implies correlation, but not vice versa!
- Slope: The variable a in the linear equation *y=a*x+b.*
- Intercept: The variable b in the linear equation *y=a*x+b.*

Bayesian Inference

Let *P(A)*, *P(B)* be the probabilities of A and B respectively. Let *P(A|B)* be the conditional probability of *A* given *B* and *P(B|A)* be the probability of *B* given *A*. Then, Bayes' theorem states:

*P(A|B)=(P(B|A) * P(A))/P(B).*

Distributions

Probability distribution is a function from the set of possible outcomes to the set of the probabilities of those outcomes.

Normal distribution

Random variables of many natural phenomena are modeled by the normal distribution. The normal distribution has the probability density:

$$f(x|\mu, \sigma^2) = \frac{e^{\frac{-(x-\mu)^2}{2\sigma^2}}}{\sqrt{2\sigma^2\pi}}$$

Where μ is the mean of the distribution and σ^2 is the variation of the distribution. The graph of the normal distribution has a shape of a bell curve, e.g. confer the graph underneath of the normal distribution with the mean 10 and the standard deviation 2.

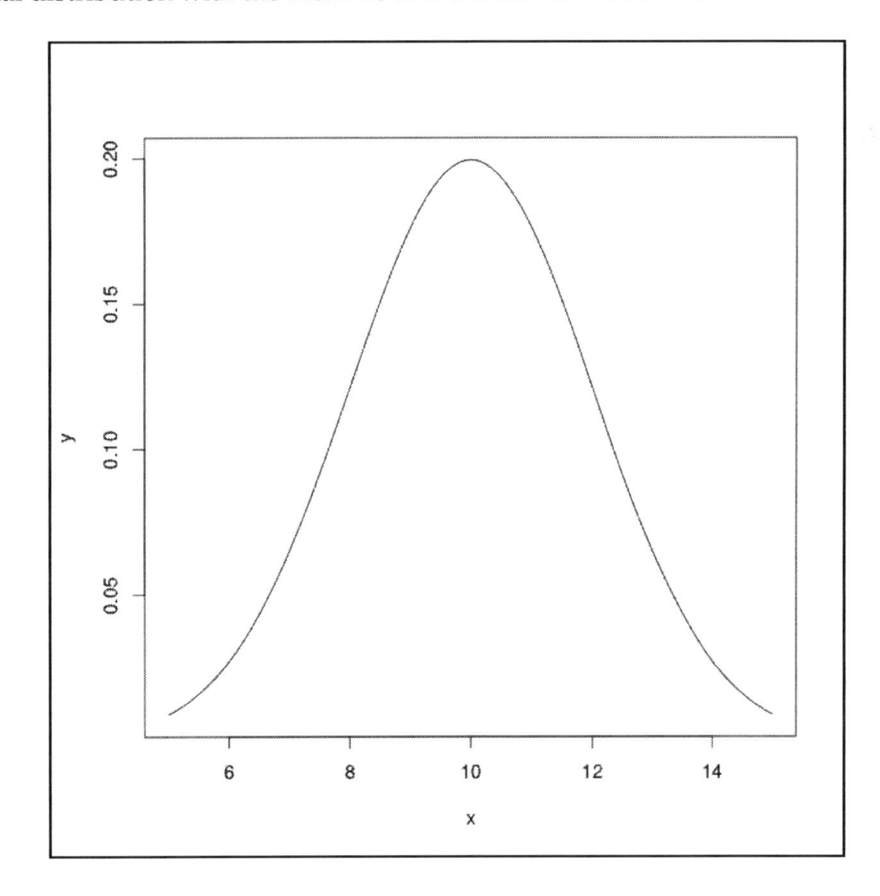

Cross-validation

Cross-validation is a method to validate an estimated hypothesis on data. In the beginning of the analysis process, the data is split into the learning data and the testing data. A hypothesis is fit to the learning data, then its actual error is measured on the testing data. This way, we can estimate how well a hypothesis may perform on the future data. Reducing the amount of learning data can also be beneficial in the end, as it reduces the chance of hypothesis over-fitting – a hypothesis being trained to a particular narrow data subset of the data.

K-fold cross-validation

Original data is partitioned randomly into the k folds. 1 fold is used for the validation, k-1 folds of data are used for hypothesis training.

A/B Testing

A/B testing is the validation of the 2 hypotheses on the data – usually on the real data. Then, the hypothesis with the better result (lower error of the estimation) is chosen as an estimator for future data.

B
R Reference

Introduction

R is a programming language with a focus on the statistical computing. For this reason, it is useful for statistics, data analysis, and data mining. R code is written in files with the suffix `.r` and can be executed with the command Rscript.

R Hello World example

A simple example in R prints one line of text.

Input:

```
source_code/appendix_b_r/example00_hello_world.r
print('Hello World!')
```

Output:

```
$ Rscript example00_hello_world.r
[1] "Hello World!"
```

Comments

Comments are not executed in R, start with the character # and end with the end of the line.

Input:

```
source_code/appendix_b_r/example01_comments.r
print("This text is printed because the print statement is executed")
#This is just a comment and will not be executed.
#print("Even commented statements are not executed.")
print("But the comment finished with the end of the line.")
print("So the 4th and 5th line of the code are executed again.")
```

Output:

```
$ Rscript example01_comments.r
[1] "This text will be printed because the print statemnt is executed"
[1] "But the comment finished with the end of the line."
[1] "So the 4th and 5th line of the code are executed again."
```

Data types

Some of the data types available in R are:

- Numeric data types: integer, numeric
- Text data types: string
- Composite data types: vector, list, data frame

Integer

The integer data type can hold only integer values:

Input:

```
source_code/appendix_b_r/example02_int.r
#Integer constants are suffixed with L.
rectangle_side_a = 10L
rectangle_side_b = 5L
rectangle_area = rectangle_side_a * rectangle_side_b
rectangle_perimeter = 2*(rectangle_side_a + rectangle_side_b)
#The command cat like print can also be used to print the output
#to the command line.
cat("Let there be a rectangle with the sides of lengths:",
```

```
rectangle_side_a, "and", rectangle_side_b, "cm.\n")
cat("Then the area of the rectangle is", rectangle_area, "cm squared.\n")
cat("The perimeter of the rectangle is", rectangle_perimeter, "cm.\n")
```

Output:

```
$ Rscript example02_int.r
Let there be a rectangle with the sides of lengths: 10 and 5 cm.
Then the area of the rectangle is 50 cm squared.
The perimeter of the rectangle is 30 cm.
```

Numeric

The numeric data type can also hold non-integer rational values.

Input:

```
source_code/appendix_b_r/example03_numeric.r
pi = 3.14159
circle_radius = 10.2
circle_perimeter = 2 * pi * circle_radius
circle_area = pi * circle_radius * circle_radius
cat("Let there be a circle with the radius", circle_radius, "cm.\n")
cat("Then the perimeter of the circle is", circle_perimeter, "cm.\n")
cat("The area of the circle is", circle_area, "cm squared.\n")
```

Output:

```
$ Rscript example03_numeric.r
Let there be a circle with the radius 10.2 cm.
Then the perimeter of the circle is 64.08844 cm.
The area of the circle is 326.851 cm squared.
```

String

A string variable can be used to store text.

Input:

```
source_code/appendix_b_r/example04_string.r
first_name = "Satoshi"
last_name = "Nakamoto"
#String concatenation is performed with the command paste.
full_name = paste(first_name, last_name, sep = " ", collapse = NULL)
cat("The invertor of Bitcoin is", full_name, ".\n")
```

Output:

```
$ Rscript example04_string.r
The invertor of Bitcoin is Satoshi-Nakamoto .
```

List and vector

Lists and vectors in R are written in brackets prefixed by the letter c. They can be used interchangeably.

Input:

```
source_code/appendix_b_r/example05_list_vector.r
some_primes = c(2, 3, 5, 7)
cat("The primes less than 10 are:", some_primes,"\n")
```

Output:

```
$ Rscript example05_list_vector.r
The primes less than 10 are: 2 3 5 7
```

Data frame

A data frame is a list of vectors of equal length.

Input:

```
source_code/appendix_b_r/example06_data_frame.r
temperatures = data.frame(
  fahrenheit = c(5,14,23,32,41,50),
  celsius = c(-15,-10,-5,0,5,10)
)
print(temperatures)
```

Output:

```
$ Rscript example06_data_frame.r
  fahrenheit celsius
1          5     -15
2         14     -10
3         23      -5
4         32       0
5         41       5
6         50      10
```

Linear regression

R is equipped with the command `lm` to fit the linear models:

Input:

```
source_code/appendix_b_r/example07_linear_regression.r
temperatures = data.frame(
    fahrenheit = c(5,14,23,32,41,50),
    celsius = c(-15,-10,-5,0,5,10)
)
model = lm(celsius ~ fahrenheit, data = temperatures)
print(model)
```

Output:

```
$ Rscript example07_linear_regression.r
Call:
lm(formula = celsius ~ fahrenheit, data = temperatures)
Coefficients:
(Intercept)   fahrenheit
   -17.7778       0.5556
```

C Python Reference

Introduction

Python is a general purpose programming and scripting language. Its simplicity and extensive libraries make it possible to develop an application quickly and compatible with the modern requirements on the technology. Python code is written in files with the suffix `.py` and can be executed with the command `python`.

Python Hello World example

A simplest program in Python prints one line of text.

Input:

```
source_code/appendix_c_python/example00_helloworld.py
print "Hello World!"
```

Output:

```
$ python example00_helloworld.py
Hello World!
```

Comments

Comments are not executed in Python, start with the character `#`, and end with the end of the line.

Input:

```
# source_code/appendix_c_python/example01_comments.py
print "This text will be printed because the print statement is executed."
#This is just a comment and will not be executed.
#print "Even commented statements are not executed."
print "But the comment finished with the end of the line."
print "So the 4th and 5th line of the code are executed again."
```

Output:

```
$ python example01_comments.py
This text will be printed because the print statement is executed
But the comment finished with the end of the line.
So the 4th and 5th line of the code are executed again.
```

Data types

Some of the data types available in Python are:

- numeric data types: int, float,
- Text data types: str
- Composite data types: tuple, list, set, dictionary.

Int

The int data type can hold only integer values.

Input:

```
# source_code/appendix_c_python/example02_int.py
rectangle_side_a = 10
rectangle_side_b = 5
rectangle_area = rectangle_side_a * rectangle_side_b
rectangle_perimeter = 2*(rectangle_side_a + rectangle_side_b)
print "Let there be a rectangle with the sides of lengths:"
print rectangle_side_a, "and", rectangle_side_b, "cm."
print "Then the area of the rectangle is", rectangle_area, "cm squared."
```

```
print "The perimeter of the rectangle is", rectangle_perimeter, "cm."
```

Output:

```
$ python example02_int.py
Let there be a rectangle with the sides of lengths: 10 and 5 cm.
Then the area of the rectangle is 50 cm squared.
The perimeter of the rectangle is 30 cm.
```

Float

The float data type can also hold non-integer rational values.

Input:

```
# source_code/appendix_c_python/example03_float.py
pi = 3.14159
circle_radius = 10.2
circle_perimeter = 2 * pi * circle_radius
circle_area = pi * circle_radius * circle_radius
print "Let there be a circle with the radius", circle_radius, "cm."
print "Then the perimeter of the circle is", circle_perimeter, "cm."
print "The area of the circle is", circle_area, "cm squared."
```

Output:

```
$ python example03_float.py
Let there be a circle with the radius 10.2 cm.
Then the perimeter of the circle is 64.088436 cm.
The area of the circle is 326.8510236 cm squared.
```

String

A string variable can be used to store text.

Input:

```
# source_code/appendix_c_python/example04_string.py
first_name = "Satoshi"
last_name = "Nakamoto"
full_name = first_name + " " + last_name
print "The inventor of Bitcoin is", full_name, "."
```

Output:

```
$ python example04_string.py
The inventor of Bitcoin is Satoshi Nakamoto .
```

Tuple

A tuple data type is analogous to a vector in mathematics. For example:

`tuple = (integer_number, float_number).`

Input:

```
# source_code/appendix_c_python/example05_tuple.py
import math

point_a = (1.2,2.5)
point_b = (5.7,4.8)
#math.sqrt computes the square root of a float number.
#math.pow computes the power of a float number.
segment_length = math.sqrt(
        math.pow(point_a[0] - point_b[0], 2) +
        math.pow(point_a[1] - point_b[1], 2))
print "Let the point A have the coordinates", point_a, "cm."
print "Let the point B have the coordinates", point_b, "cm."
print "Then the length of the line segment AB is", segment_length, "cm."
```

Output:

```
$ python example05_tuple.py
Let the point A have the coordinates (1.2, 2.5) cm.
Let the point B have the coordinates (5.7, 4.8) cm.
Then the length of the line segment AB is 5.0537115074 cm.
```

List

A list in Python is an ordered set of values.

Input:

```
# source_code/appendix_c_python/example06_list.py
some_primes = [2, 3]
some_primes.append(5)
some_primes.append(7)
print "The primes less than 10 are:", some_primes
```

Output:

```
$ python example06_list.py
The primes less than 10 are: [2, 3, 5, 7]
```

Set

A set in Python is a non-ordered mathematical set of values.

Input:

```
# source_code/appendix_c_python/example07_set.py
from sets import Set

boys = Set(['Adam', 'Samuel', 'Benjamin'])
girls = Set(['Eva', 'Mary'])
teenagers = Set(['Samuel', 'Benjamin', 'Mary'])
print 'Adam' in boys
print 'Jane' in girls
girls.add('Jane')
print 'Jane' in girls
teenage_girls = teenagers & girls #intersection
mixed = boys | girls #union
non_teenage_girls = girls - teenage_girls #difference
print teenage_girls
print mixed
print non_teenage_girls
```

Output:

```
$ python example07_set.py
True
False
True
Set(['Mary'])
Set(['Benjamin', 'Adam', 'Jane', 'Eva', 'Samuel', 'Mary'])
Set(['Jane', 'Eva'])
```

Dictionary

A dictionary is a data structure that can store values by their keys.

Input:

```
# source_code/appendix_c_python/example08_dictionary.py
dictionary_names_heights = {}
dictionary_names_heights['Adam'] = 180.
dictionary_names_heights['Benjamin'] = 187
dictionary_names_heights['Eva'] = 169
print 'The height of Eva is', dictionary_names_heights['Eva'], 'cm.'
```

Output:

```
$ python example08_dictionary.py
The height of Eva is 169 cm.
```

Flow control

Conditionals, We can make certain amount of the code to be executed only upon a certain condition met using the if statement. If the condition is not met, then we can execute the code following the else statement. If the first condition is not met, we can set the next condition for the code to be executed using the elif statement.

Input:

```
# source_code/appendix_c_python/example09_if_else_elif.py
x = 10
if x == 10:
        print 'The variable x is equal to 10.'

if x > 20:
        print 'The variable x is greater than 20.'
else:
        print 'The variable x is not greater than 20.'

if x > 10:
        print 'The variable x is greater than 10.'
elif x > 5:
        print 'The variable x is not greater than 10, but greater ' +
              'than 5.'
else:
        print 'The variable x is not greater than 5 or 10.'
```

Output:

```
$ python example09_if_else_elif.py
The variable x is equal to 10.
The variable x is not greater than 20.
The variable x is not greater than 10, but greater than 5.
```

For loop

For loop enables the iteration through every element in some set of elements, e.g. range, python set, list.

For loop on range

Input:

```
source_code/appendix_c_python/example10_for_loop_range.py
print "The first 5 positive integers are:"
for i in range(1,6):
       print i
```

Output:

```
$ python example10_for_loop_range.py
The first 5 positive integers are:
1
2
3
4
5
```

For loop on list

Input:

```
source_code/appendix_c_python/example11_for_loop_list.py
primes = [2, 3, 5, 7, 11, 13]
print 'The first', len(primes), 'primes are:'
for prime in primes:
       print prime
```

Output:

```
$ python example11_for_loop_list.py
The first 6 primes are:
2
3
5
7
11
13
```

Break and continue

For loops can be exited earlier with the statement break. The rest of the cycle in the for loop can be skipped with the statement continue.

Input:

```
source_code/appendix_c_python/example12_break_continue.py
for i in range(0,10):
        if i % 2 == 1: #remainder from the division by 2
                continue
        print 'The number', i, 'is divisible by 2.'

for j in range(20,100):
        print j
        if j > 22:
                break;
```

Output:

```
$ python example12_break_continue.py
The number 0 is divisible by 2.
The number 2 is divisible by 2.
The number 4 is divisible by 2.
The number 6 is divisible by 2.
The number 8 is divisible by 2.
20
21
22
23
```

Functions

Python supports the definition of the functions which is a good way to define a piece of code that is executed at multiple places in the program. A function is defined using the keyword def.

Input:

```
source_code/appendix_c_python/example13_function.py
def rectangle_perimeter(a, b):
        return 2 * (a + b)

print 'Let a rectangle have its sides 2 and 3 units long.'
print 'Then its perimeter is', rectangle_perimeter(2, 3), 'units.'
print 'Let a rectangle have its sides 4 and 5 units long.'
print 'Then its perimeter is', rectangle_perimeter(4, 5), 'units.'
```

Output:

```
$ python example13_function.py
Let a rectangle have its sides 2 and 3 units long.
Then its perimeter is 10 units.
Let a rectangle have its sides 4 and 5 units long.
Then its perimeter is 18 units.
```

Program arguments

A program can be passed arguments from the command line.

Input:

```
source_code/appendix_c_python/example14_arguments.py
#Import the system library in order to use the argument list.
import sys

print 'The number of the arguments given is', len(sys.argv),'arguments.'
print 'The argument list is ', sys.argv, '.'
```

Output:

```
$ python example14_arguments.py arg1 110
The number of the arguments given is 3 arguments.
The argument list is  ['example14_arguments.py', 'arg1', '110'] .
```

Reading and writing the file

The following program will write two lines into the file `test.txt`, then read them and finally print them to the output.

Input:

```
# source_code/appendix_c_python/example15_file.py
#write to the file with the name "test.txt"
file = open("test.txt","w")
file.write("first line\n")
file.write("second line")
file.close()

#read the file
file = open("test.txt","r")
print file.read()
```

Output:

```
$ python example15_file.py
first line
second line
```

D

Glossary of Algorithms and Methods in Data Science

- **k-Nearest Neighbors algorithm:** An algorithm that estimates an unknown data item to be like the majority of the k-closest neighbors to that item.
- **Naive Bayes classifier**: A way to classify a data item using Bayes' theorem about the conditional probabilities, $P(A|B)=(P(B|A) * P(A))/P(B)$, and in addition, assuming the independence between the given variables in the data.
- **Decision Tree**: A model classifying a data item into one of the classes at the leaf node, based on the matching properties between the branches on the tree and the actual data item.
- **Random Decision Tree**: A decision tree in which every branch is formed using only a random subset of the available variables during its construction.
- **Random Forest**: An ensemble of random decision trees constructed on the random subset of the data with the replacement, where a data item is classified to the class with the majority vote from its trees.
- **K-means algorithm**: The clustering algorithm that divides the dataset into the k groups such that the members in the group are as similar possible, that is, closest to each other.
- **Regression analysis**: A method of the estimation of the unknown parameters in a functional model predicting the output variable from the input variables, for example, to estimate a and b in the linear model y=a*x+b.

- **Time series analysis**: The analysis of data dependent on time; it mainly includes the analysis of trend and seasonality.
- **Support vector machines**: A classification algorithm that finds the hyperplane that divides the training data into the given classes. This division by the hyperplane is then used to classify the data further.
- **Principal component analysis**: The preprocessing of the individual components of the given data in order to achieve better accuracy, for example, rescaling of the variables in the input vector depending on how much impact they have on the end result.
- **Text mining**: The search and extraction of text and its possible conversion to numerical data used for data analysis.
- **Neural networks**: A machine learning algorithm consisting of a network of simple classifiers making decisions based on the input or the results of the other classifiers in the network.
- **Deep learning**: The ability of a neural network to improve its learning process.
- **A priori association rules**: The rules that can be observed in the training data and, based on which, a classification of the future data can be made.
- **PageRank**: A search algorithm that assigns the greatest relevance to the search result that has the greatest number of incoming web links from the most relevant search results on a given search term. In mathematical terms, PageRank calculates a certain eigenvector representing these measures of relevance.
- **Ensemble learning**: A method of learning where different learning algorithms are used to make a final conclusion.
- **Bagging**: A method of classifying a data item by the majority vote of the classifiers trained on the random subsets of the training data.
- **Genetic algorithms**: Machine learning algorithms inspired by the genetic processes, for example, an evolution where classifiers with the best accuracy are trained further.
- **Inductive inference**: A machine learning method learning the rules that produced the actual data.
- **Bayesian networks**: A graph model representing random variables with their conditional dependencies.
- **Singular value decomposition**: A factorization of a matrix, a generalization of eigen decomposition, used in least squares methods.
- **Boosting**: A machine learning meta algorithm decreasing the variance in the estimation by making a prediction based on the ensembles of the classifiers.
- **Expectation maximization**: An iterative method to search the parameters in the model that maximize the accuracy of the prediction of the model.

Index

S

T

Made in the USA
Middletown, DE
20 November 2017